MINDFULNESS
COLORING BOOK
ANIMAL, NATURE AND MAGIC DREAM DESIGNS

TEST YOUR COLOR

TEST YOUR COLOR

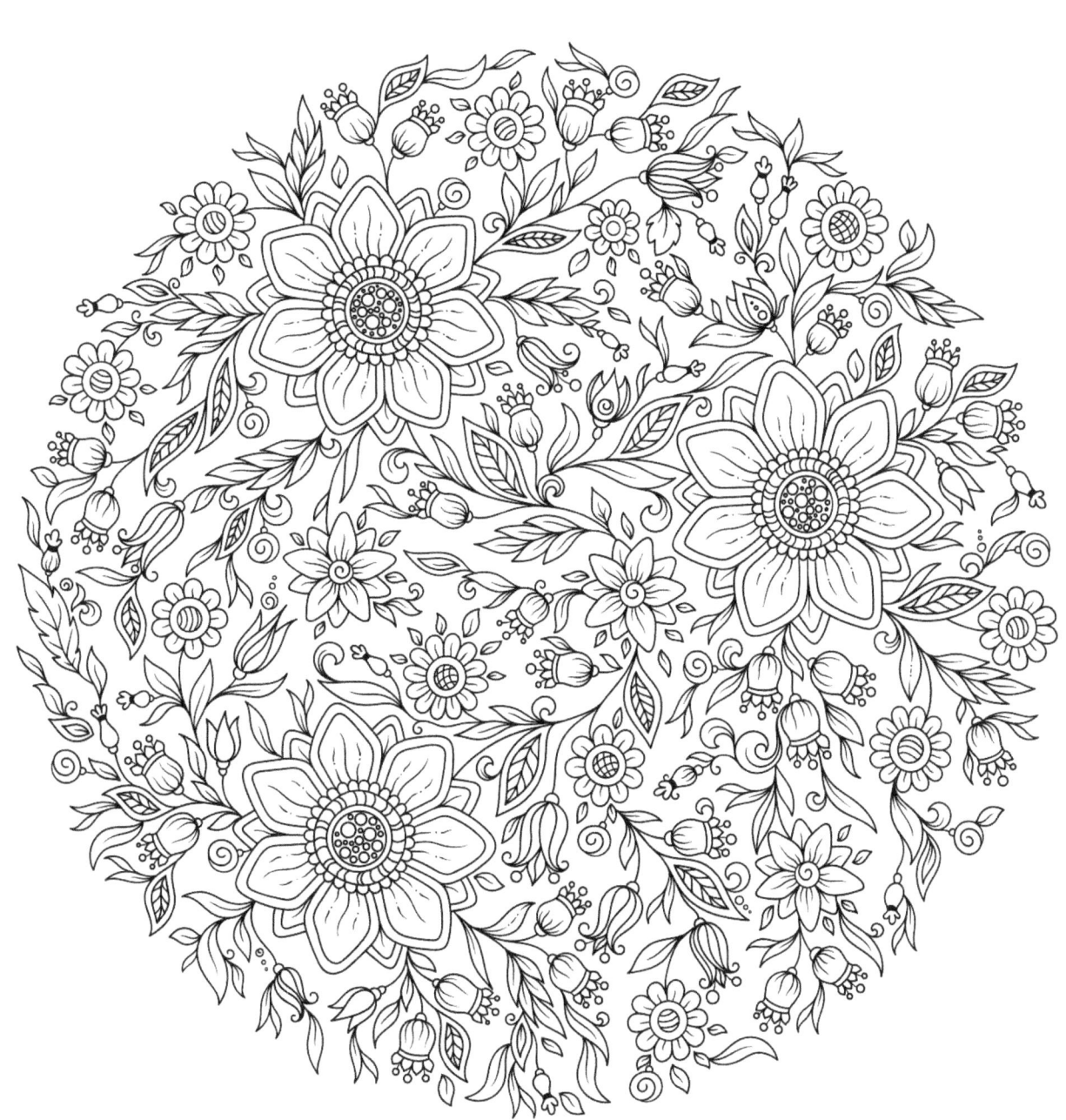

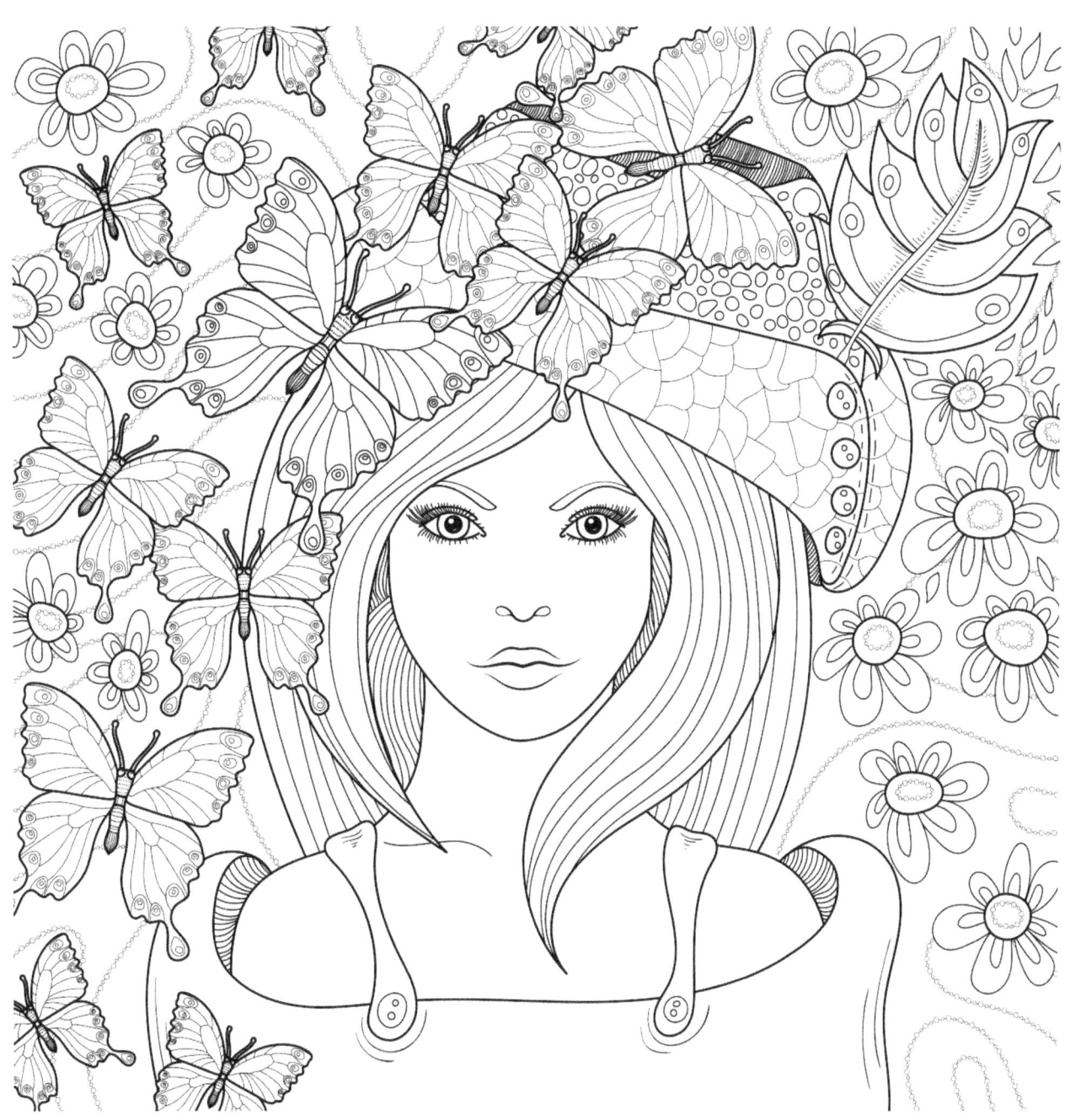

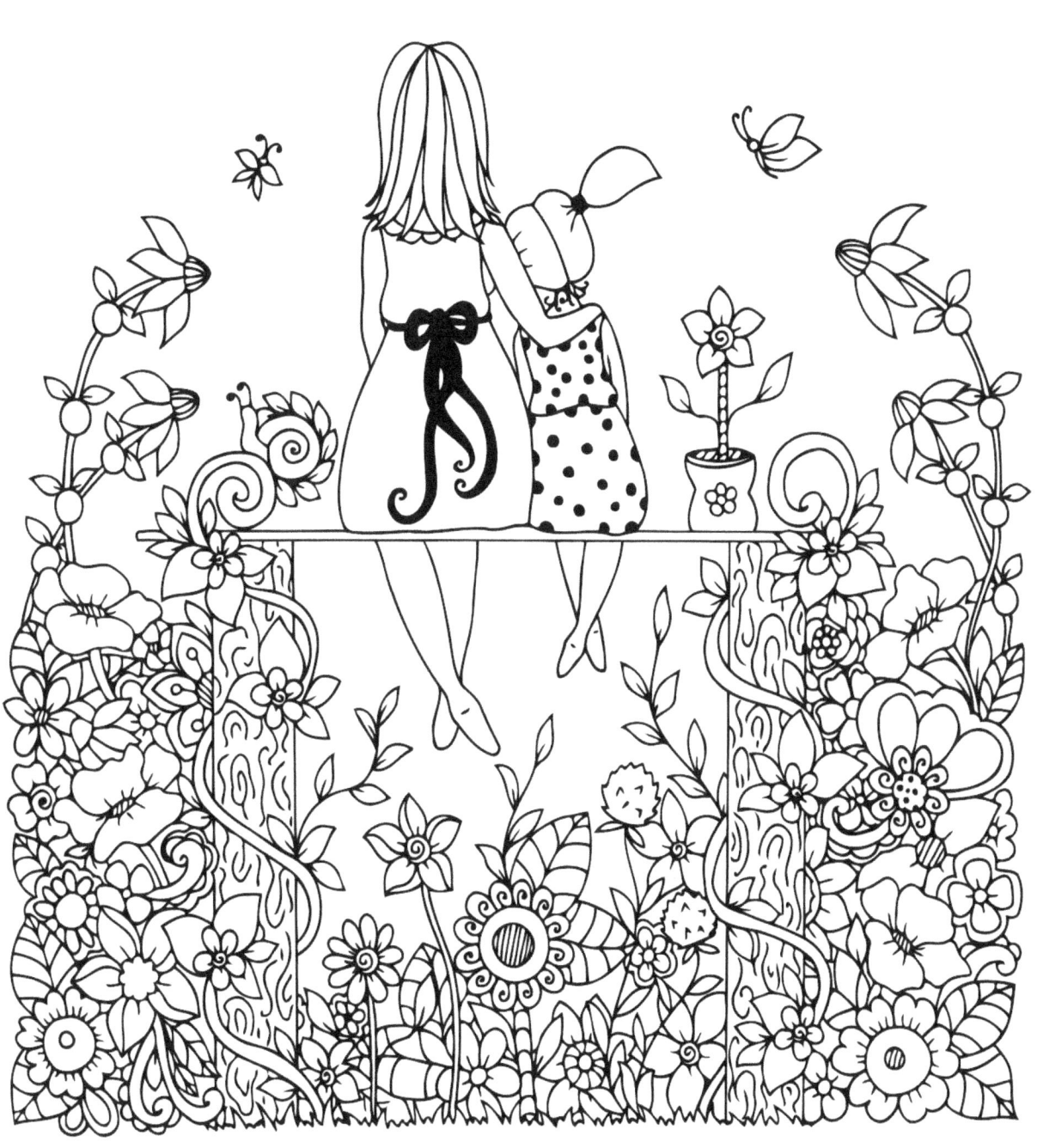

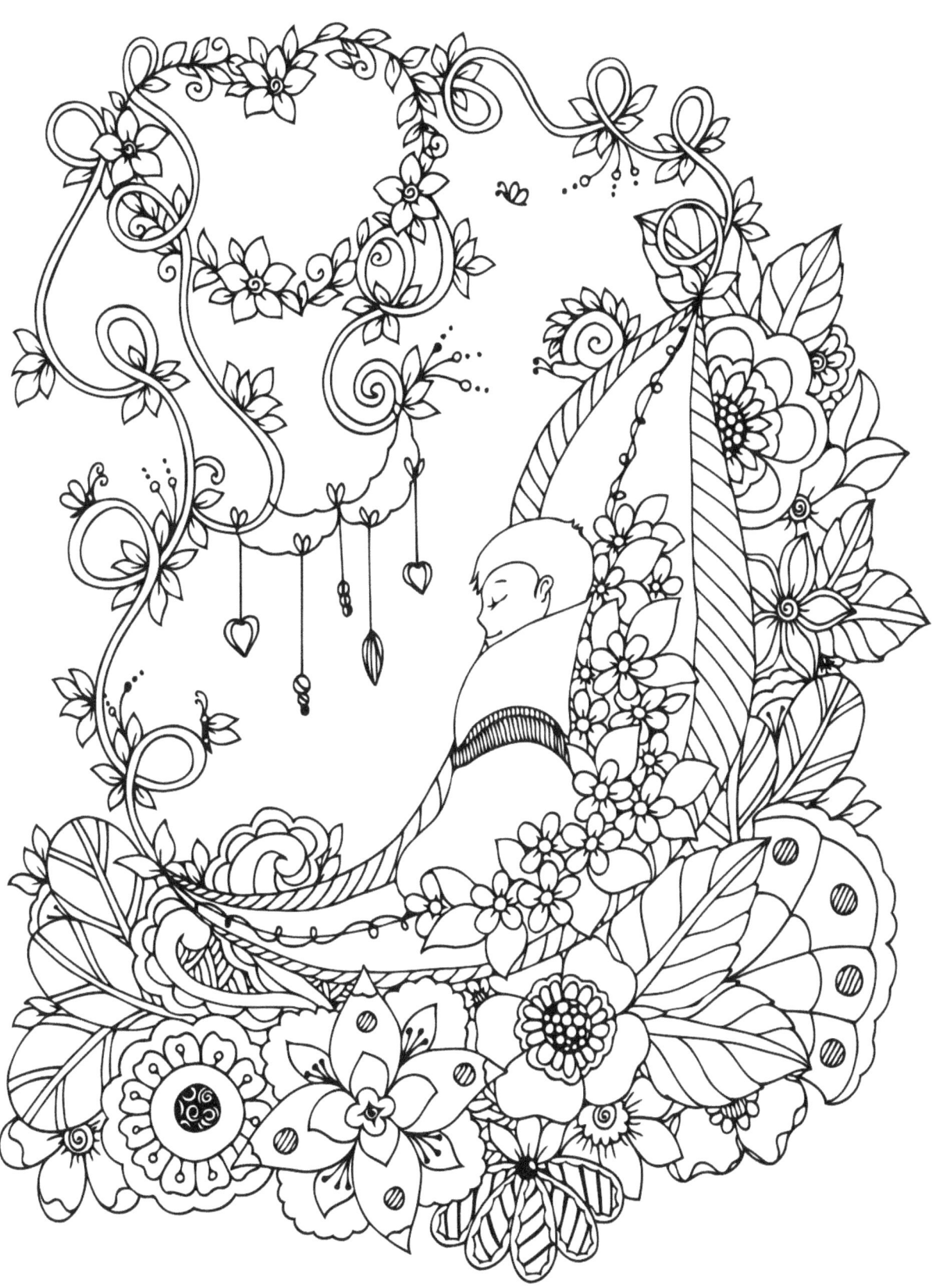

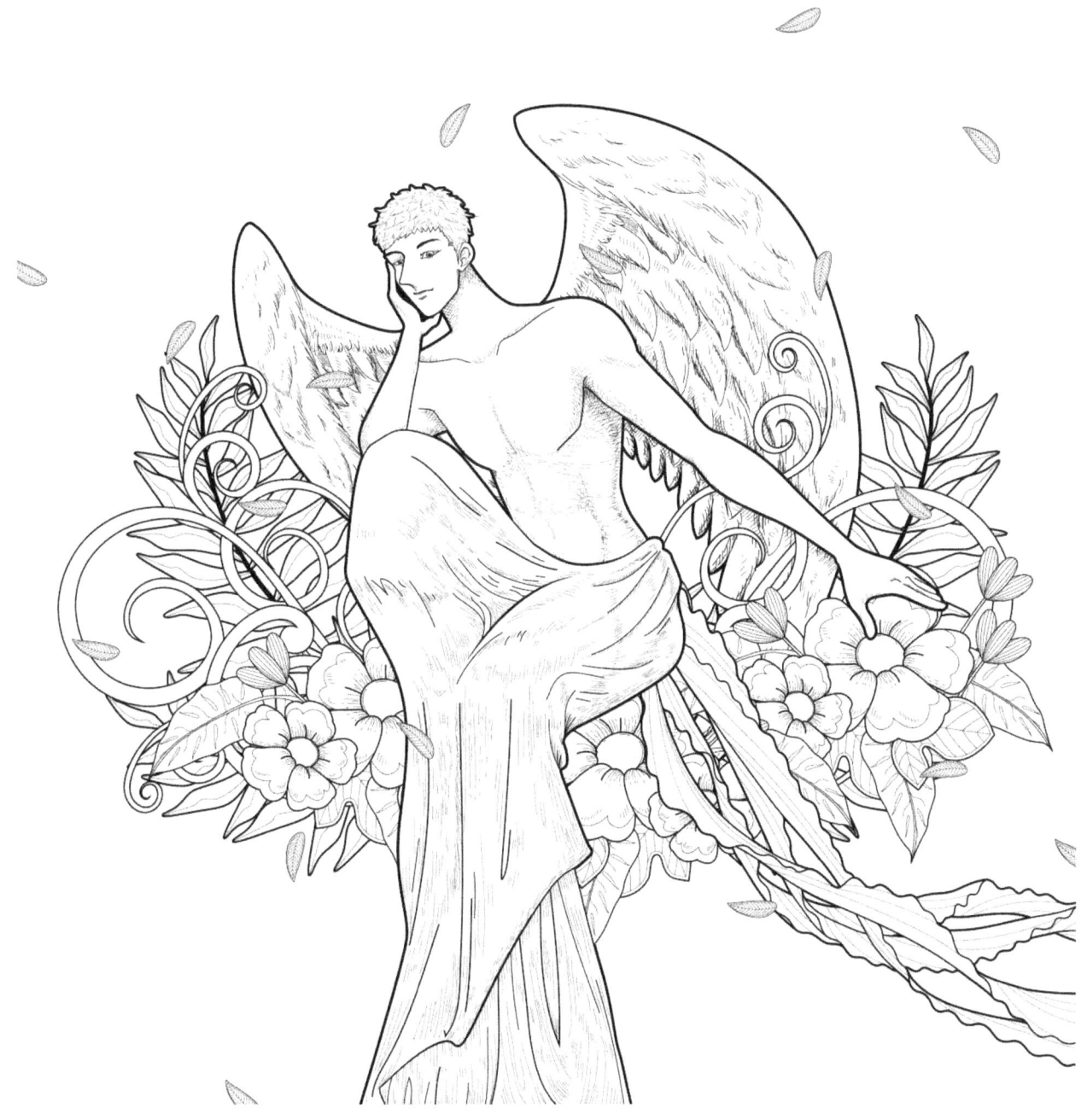

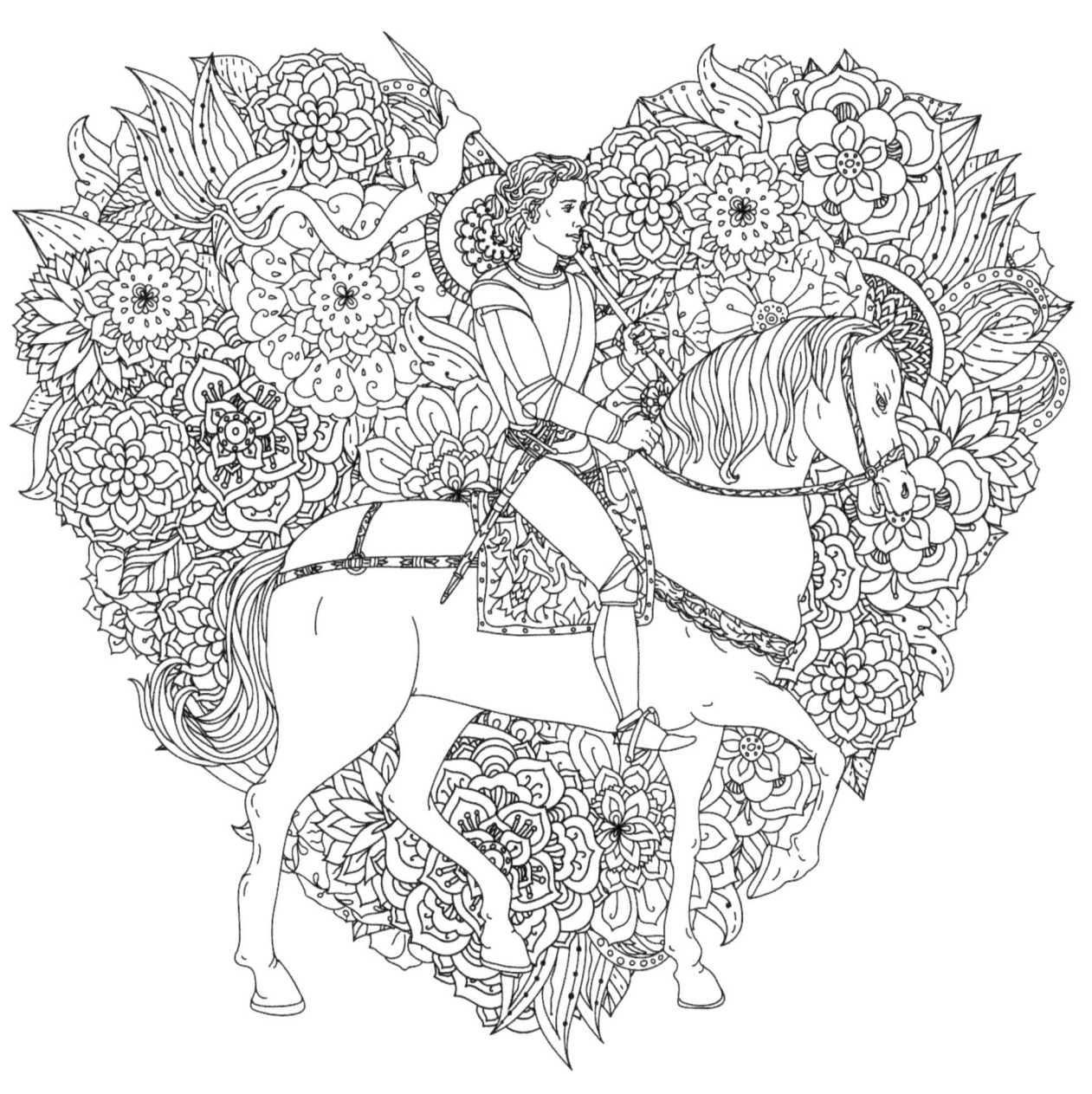

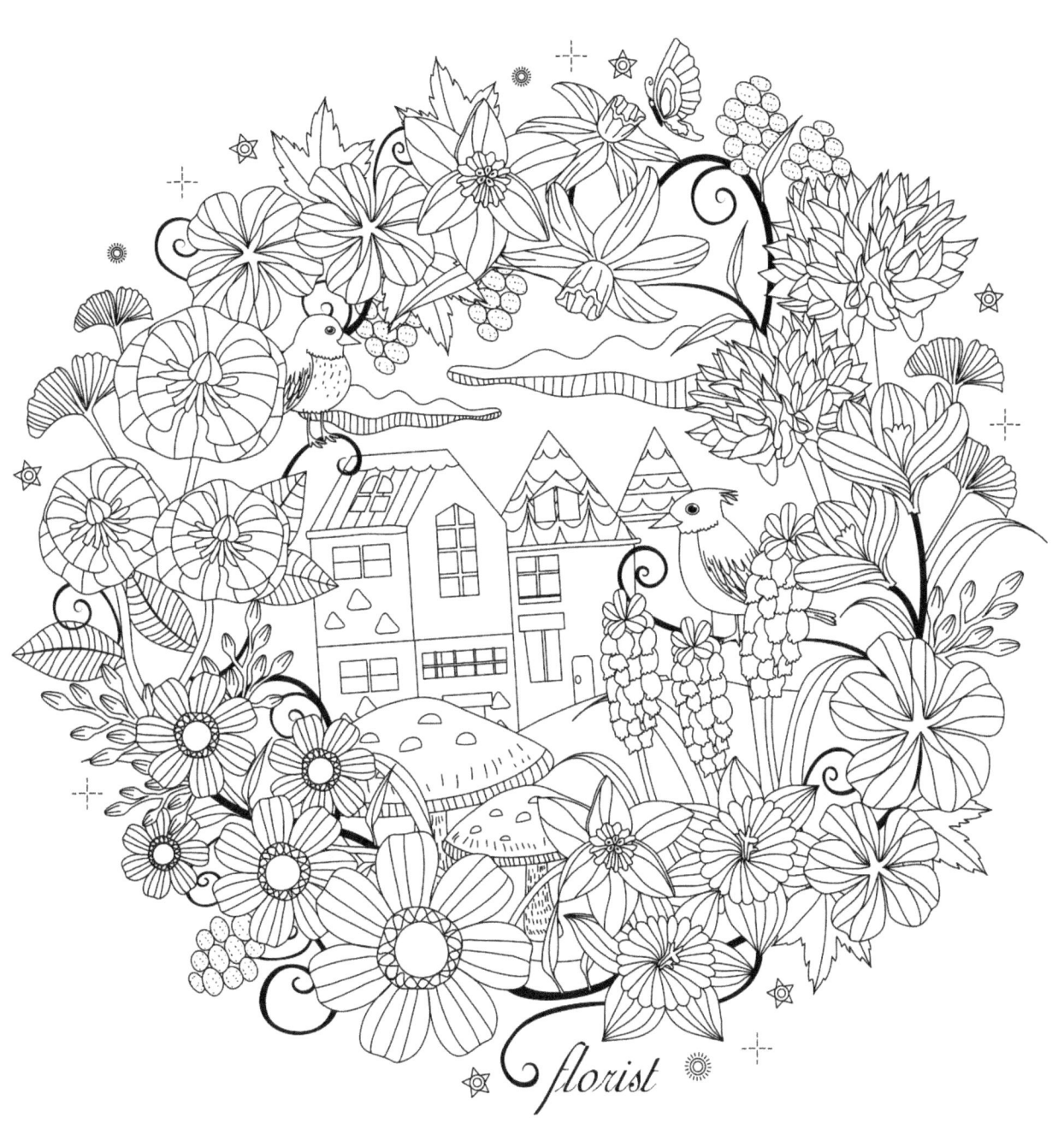

florist

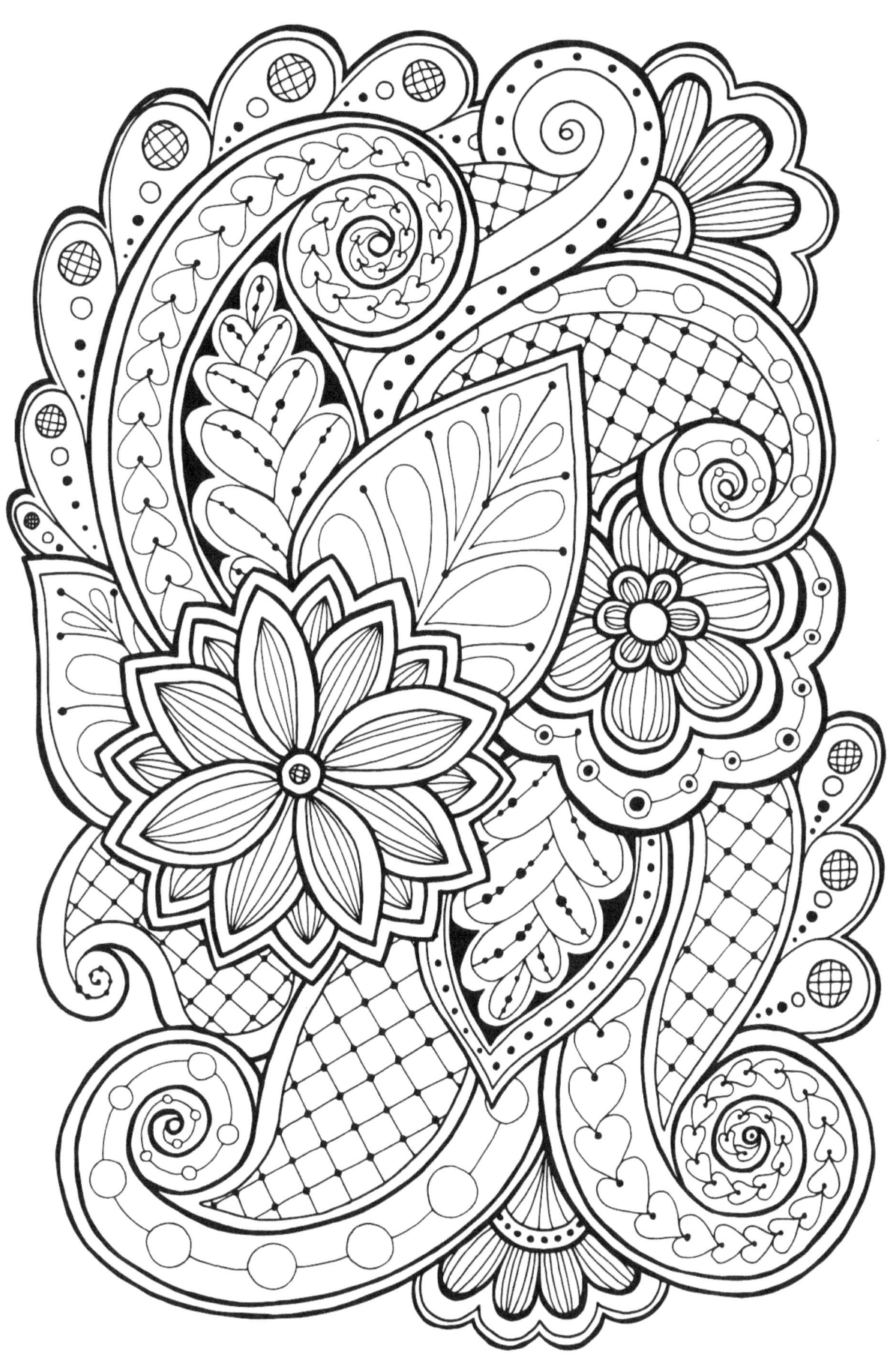

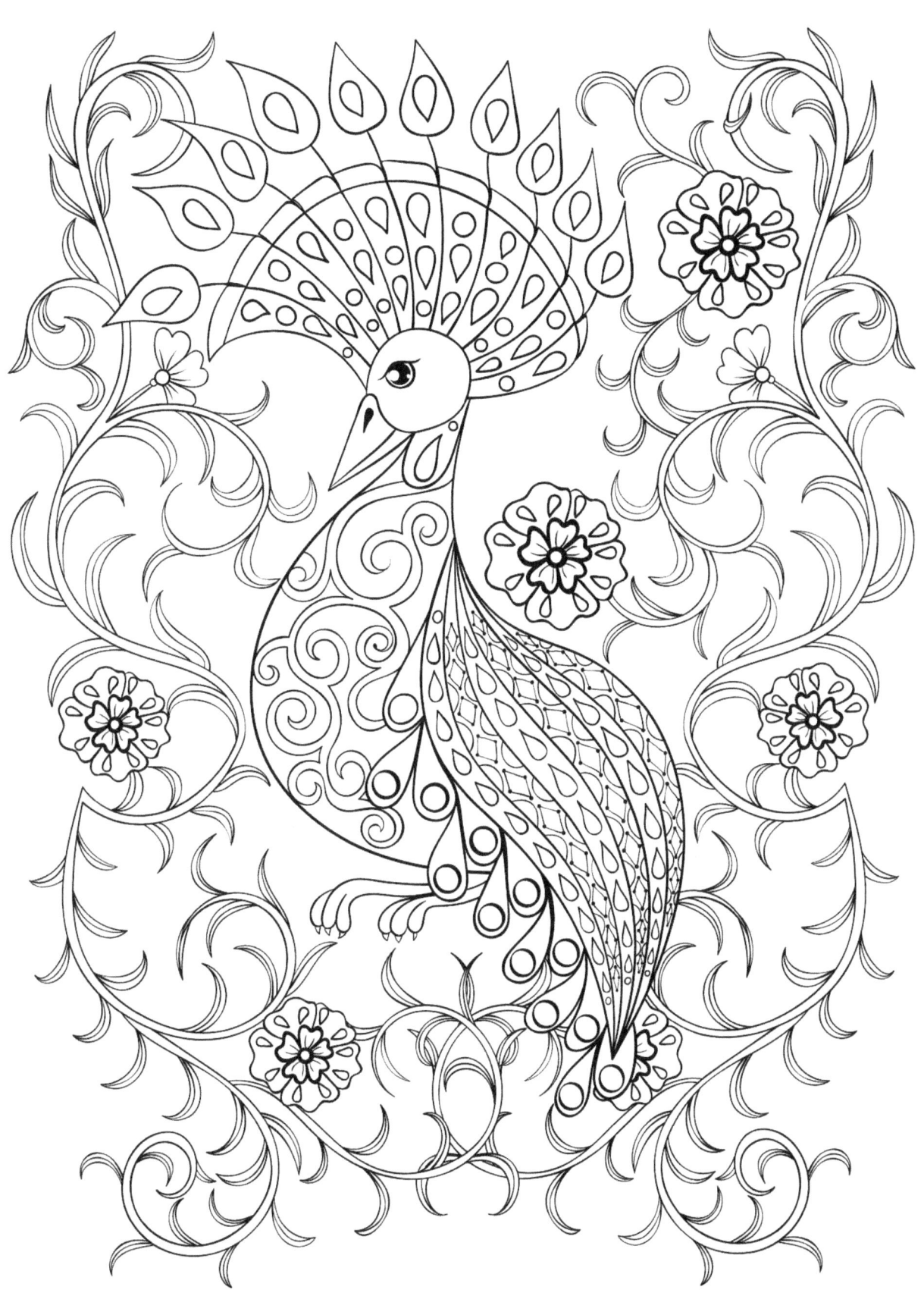

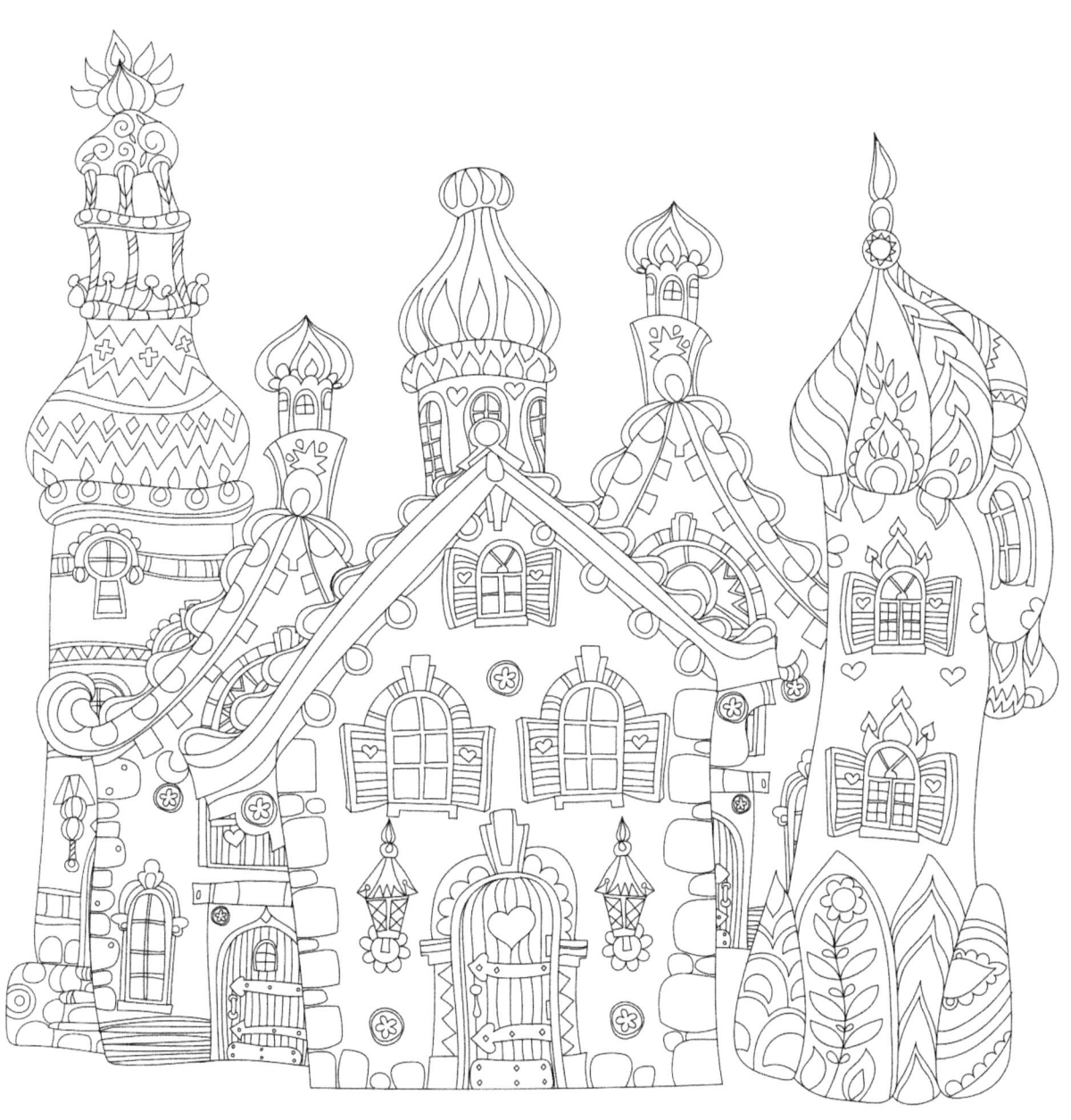

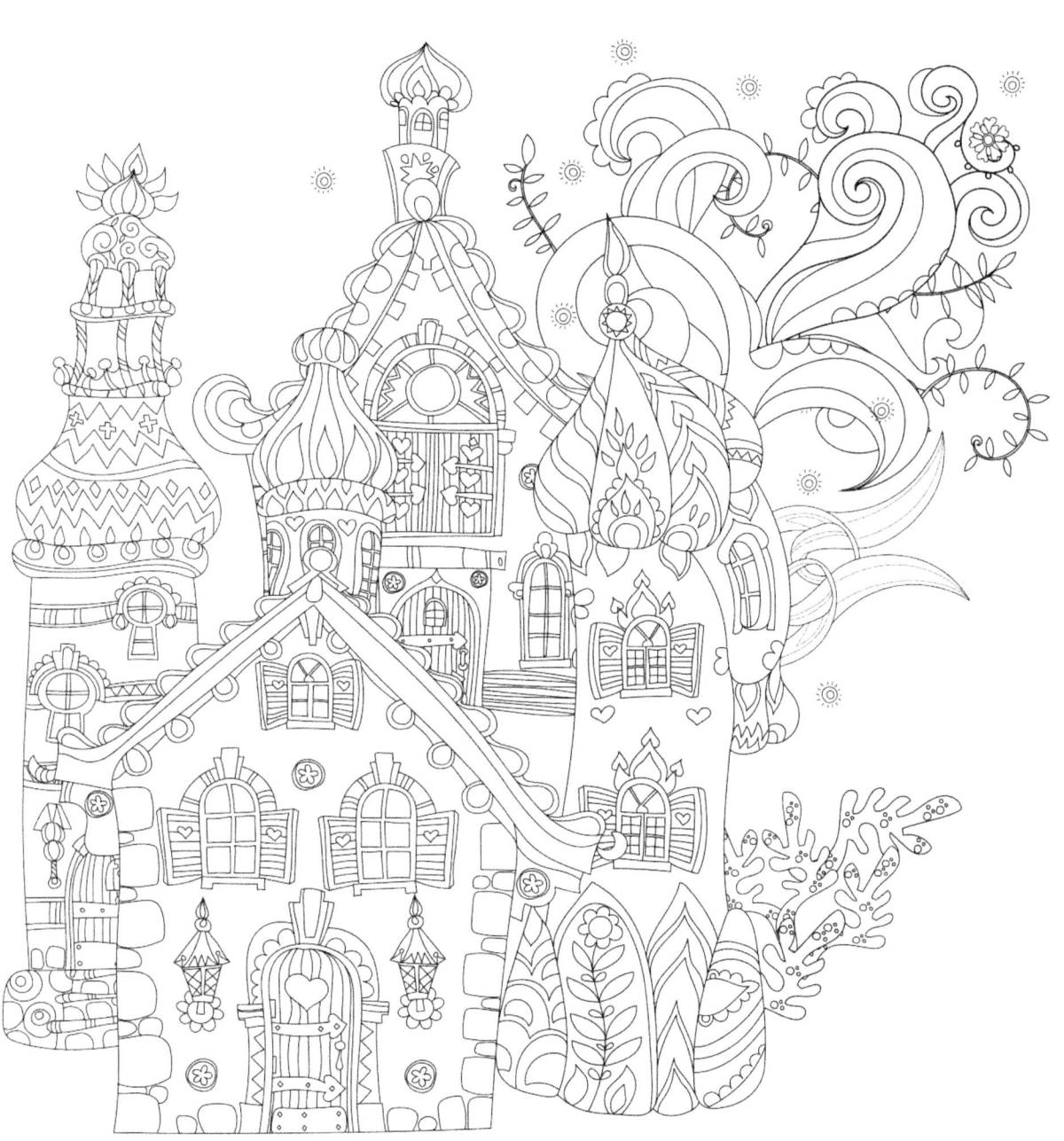

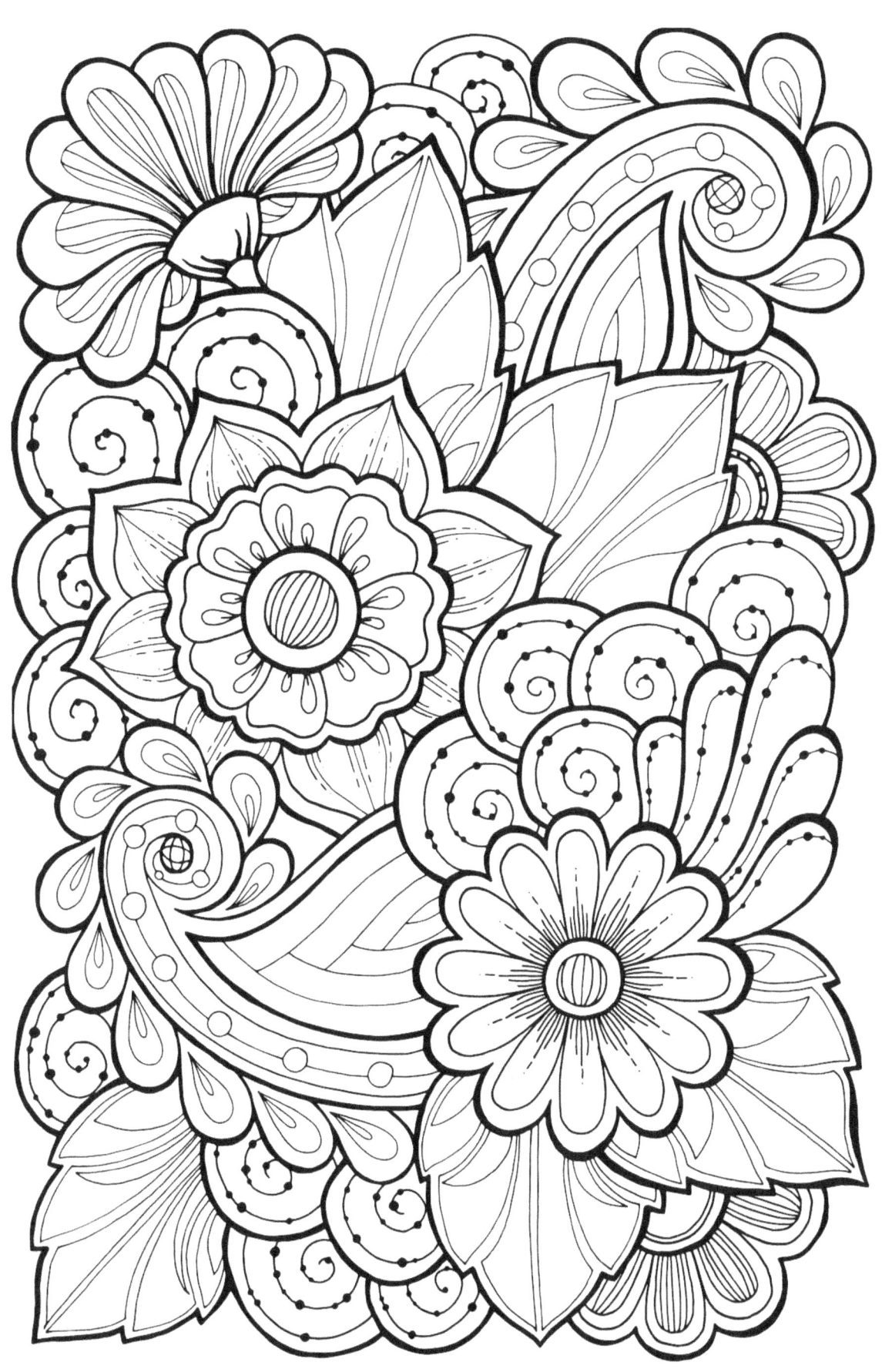

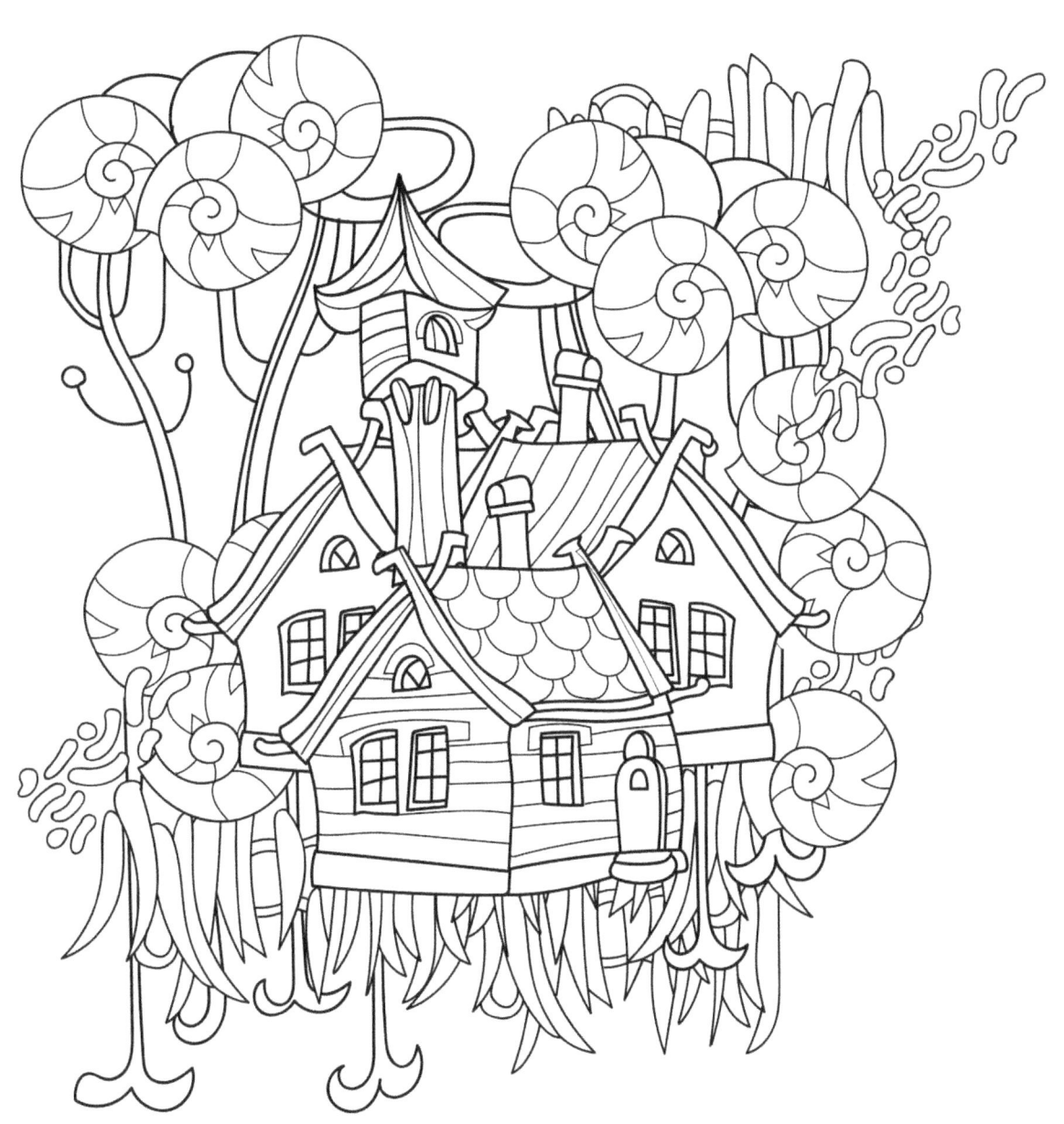

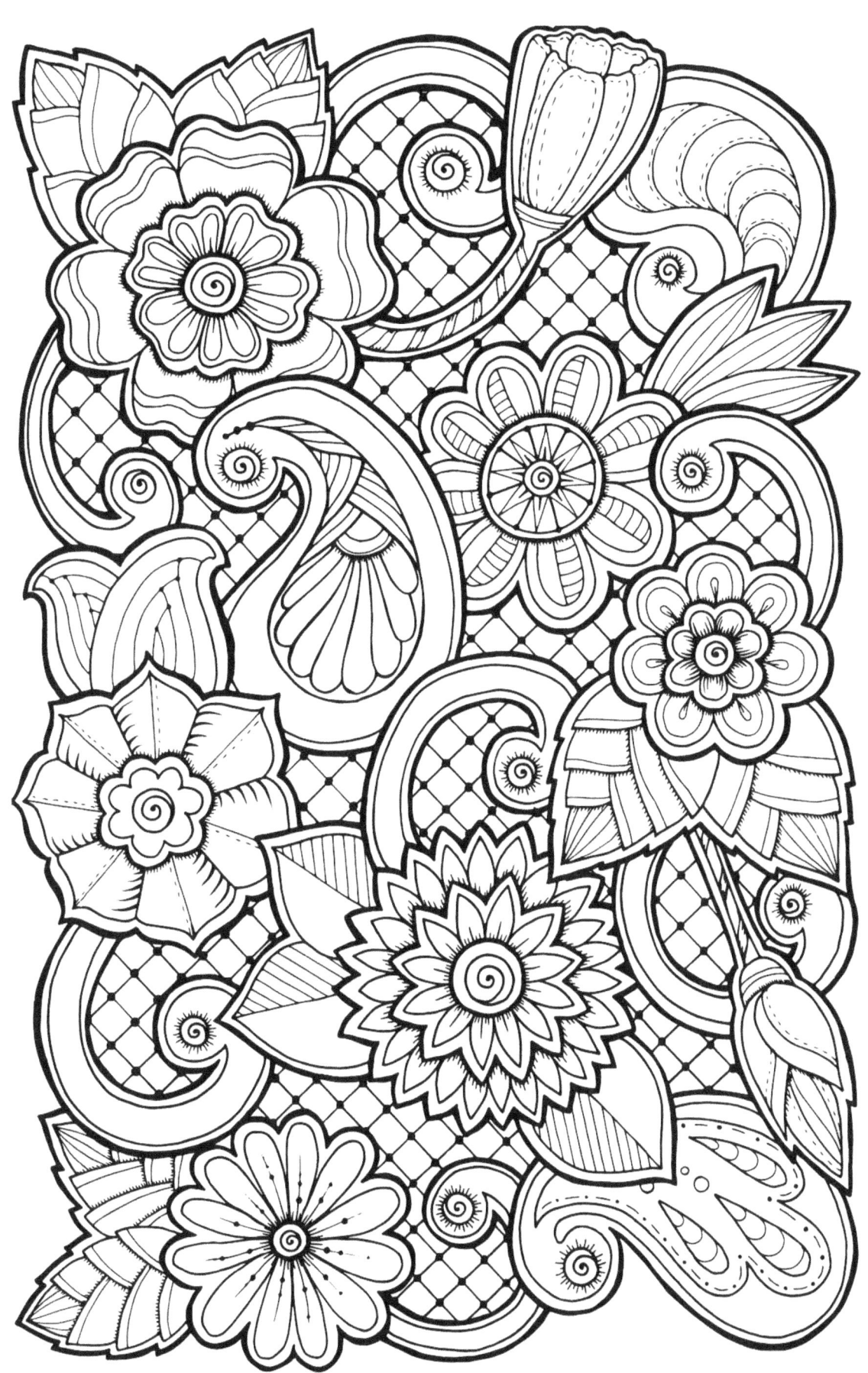

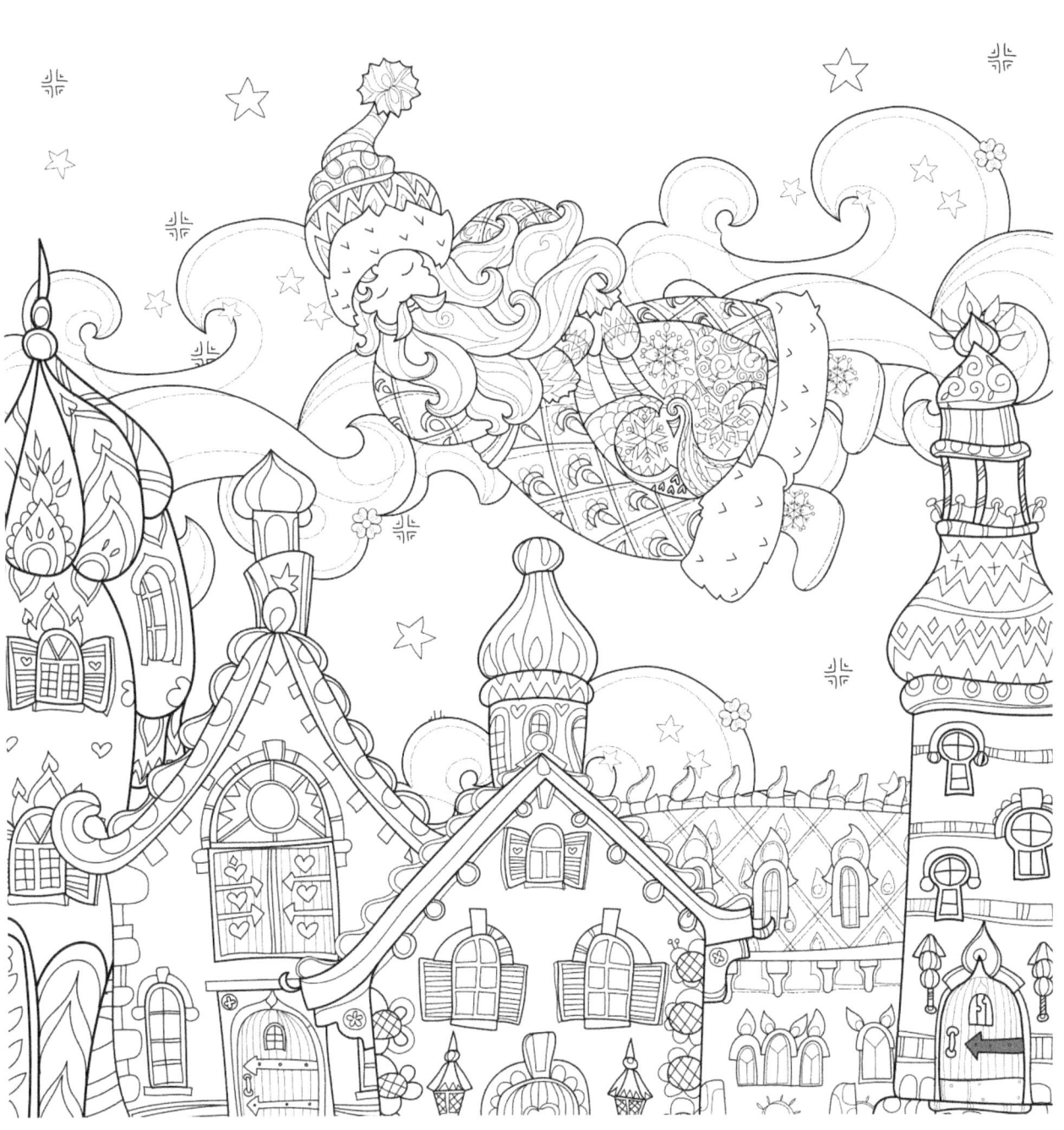

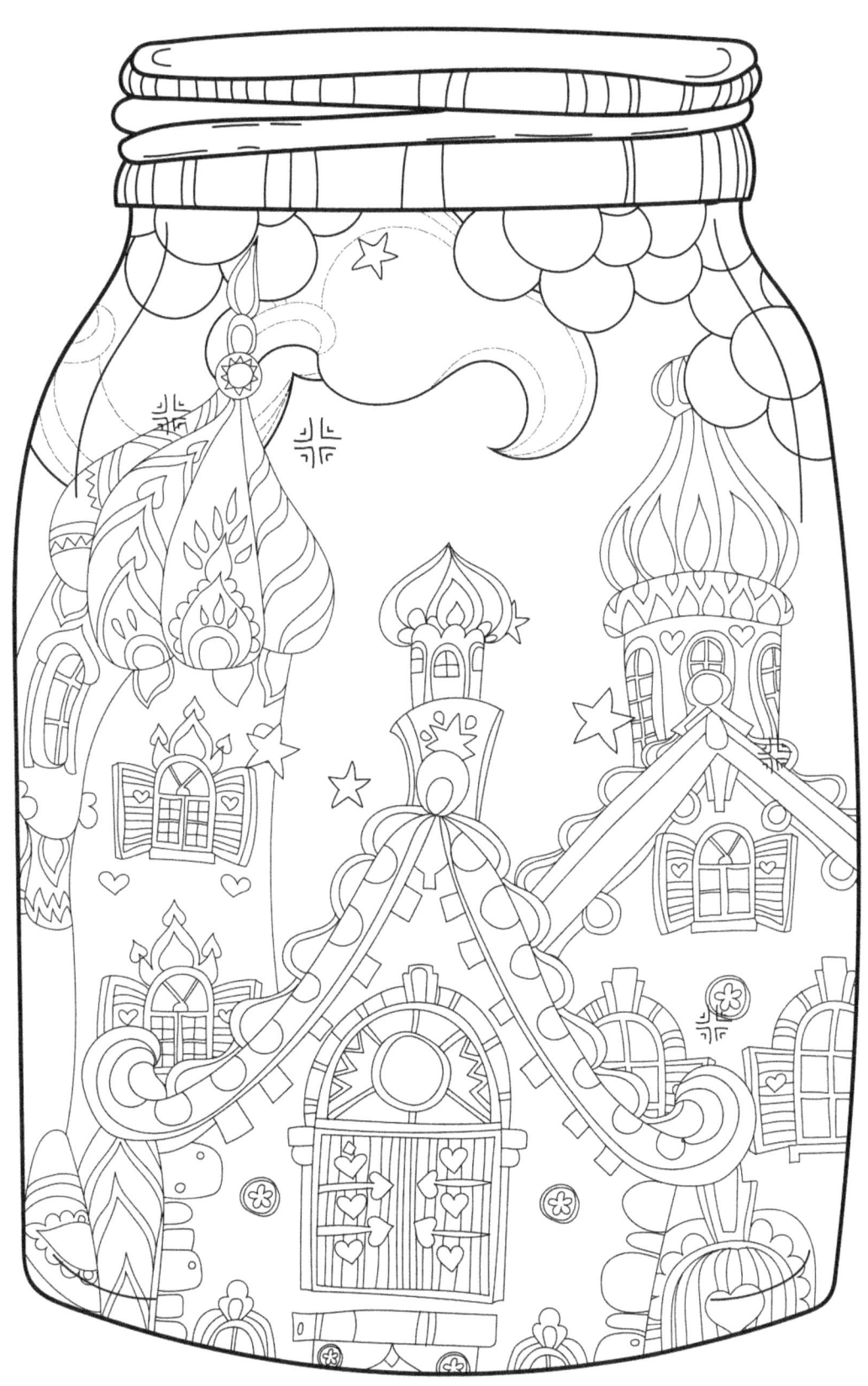

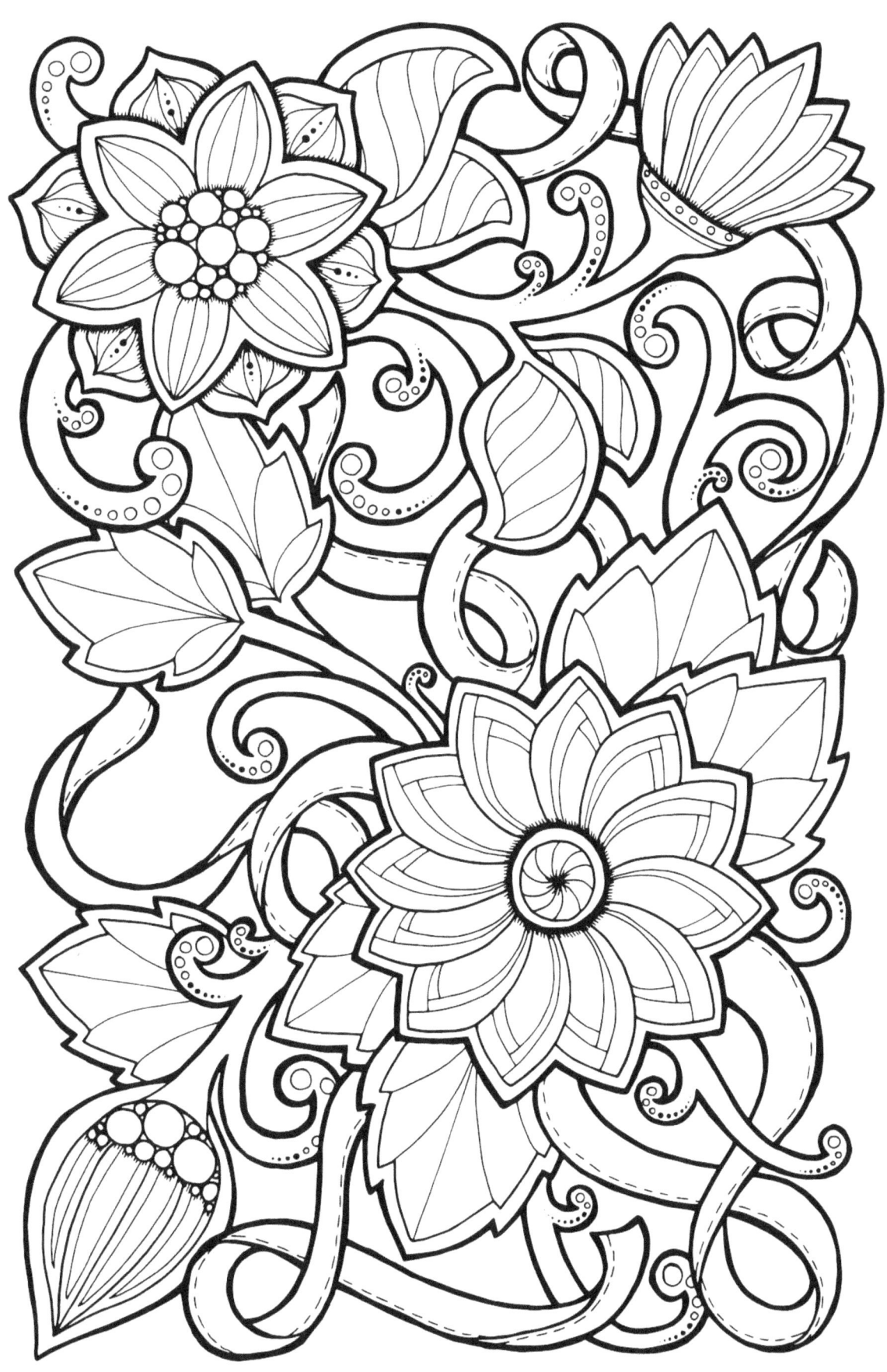

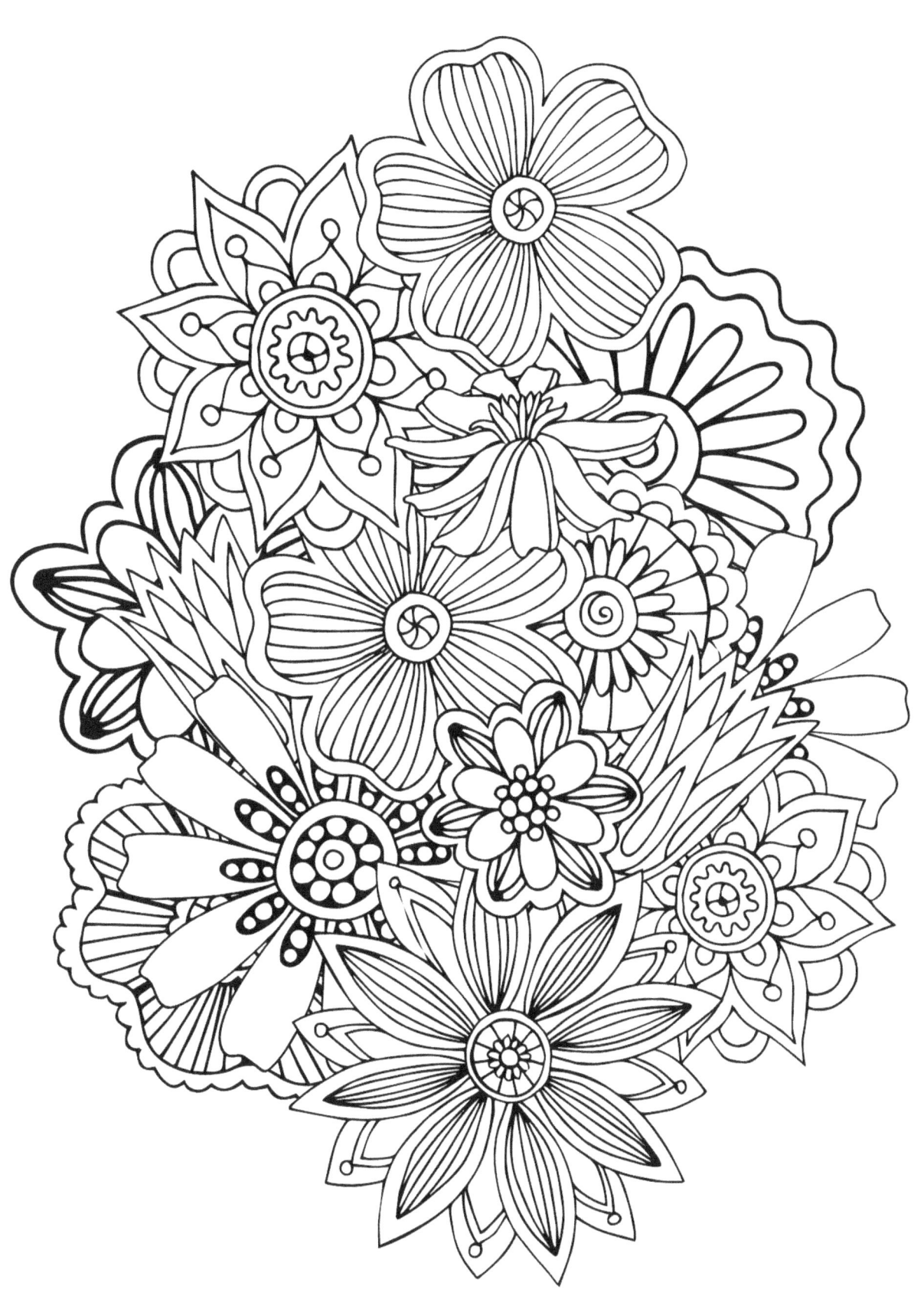

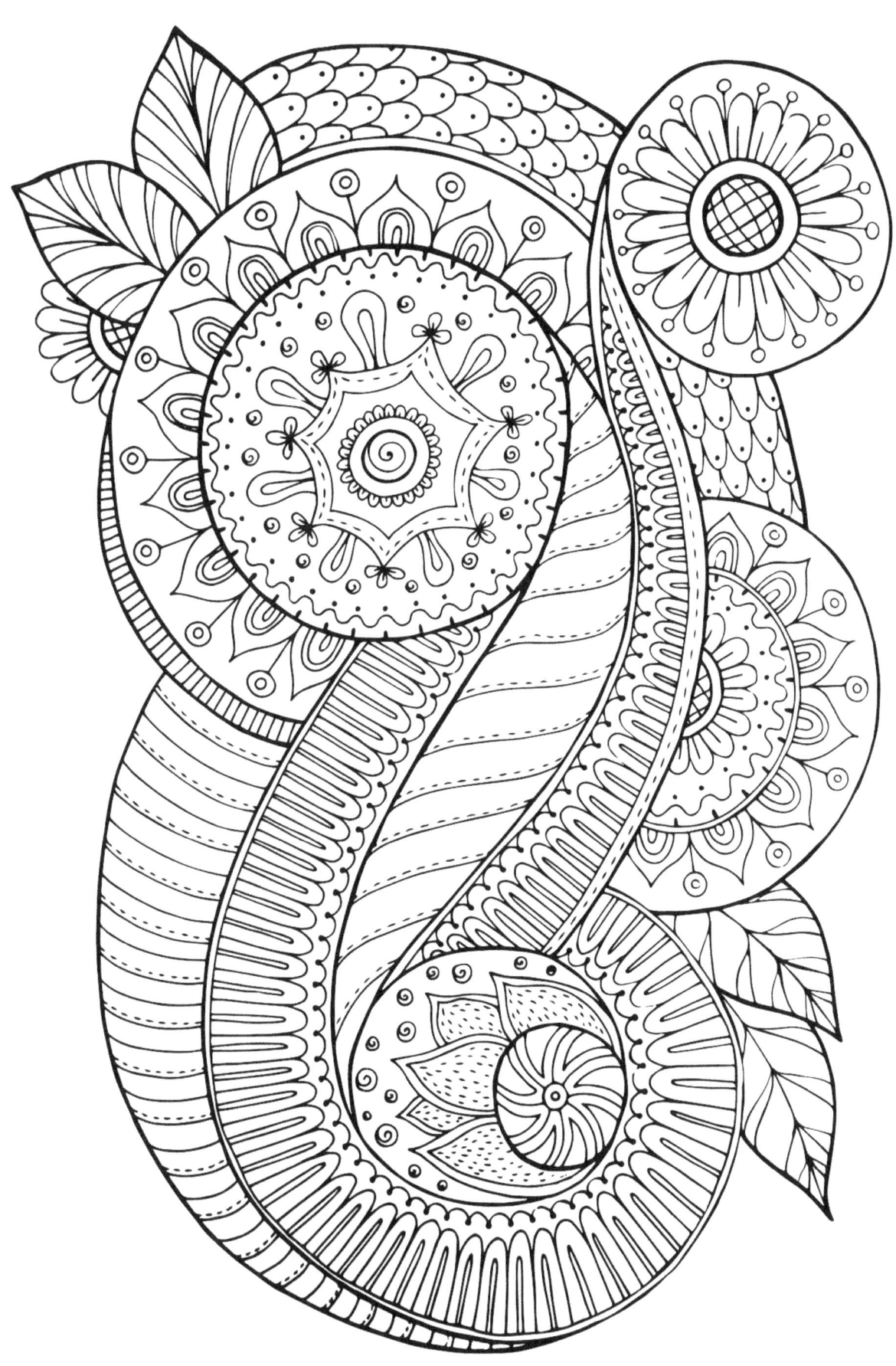

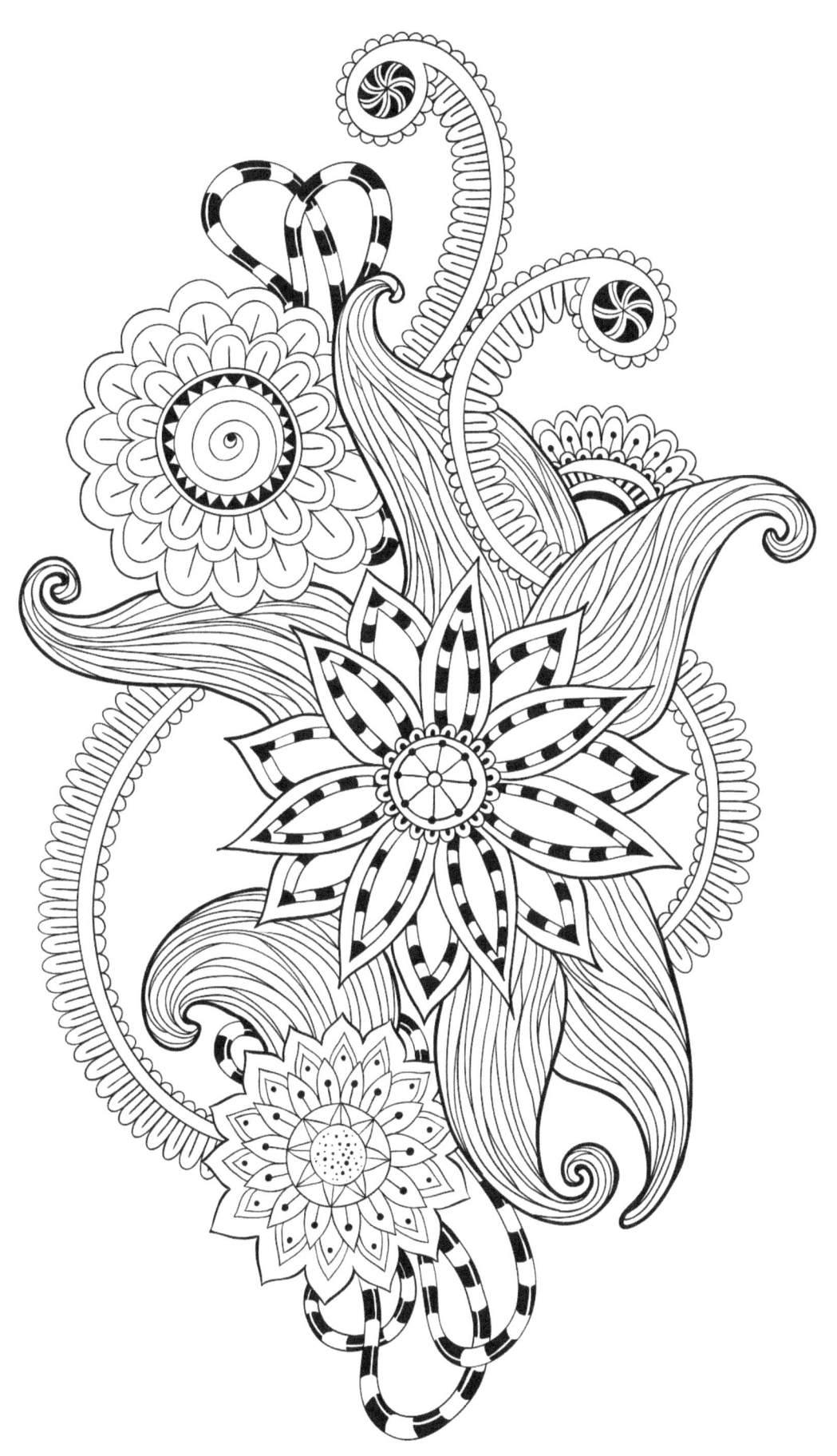

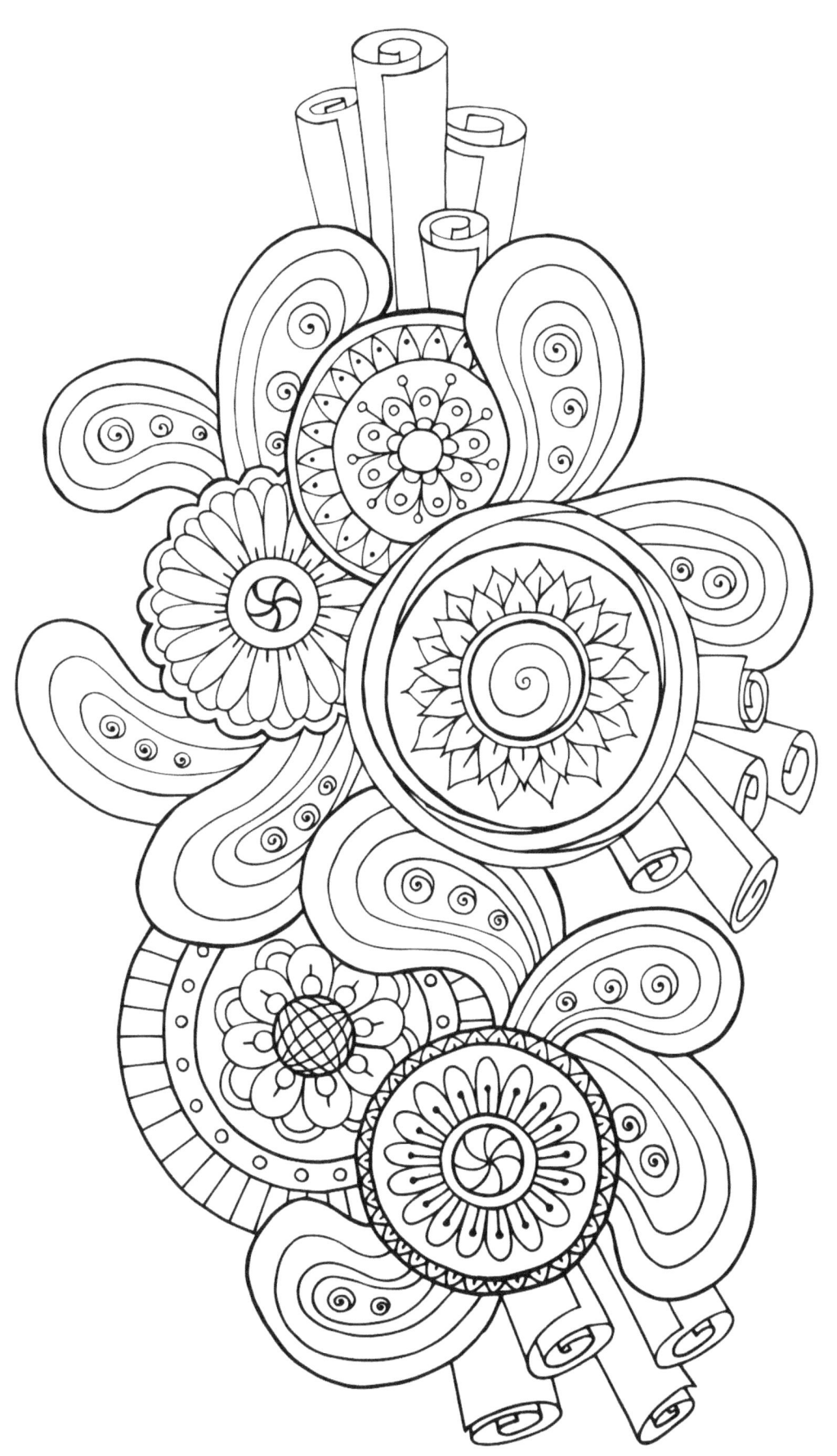

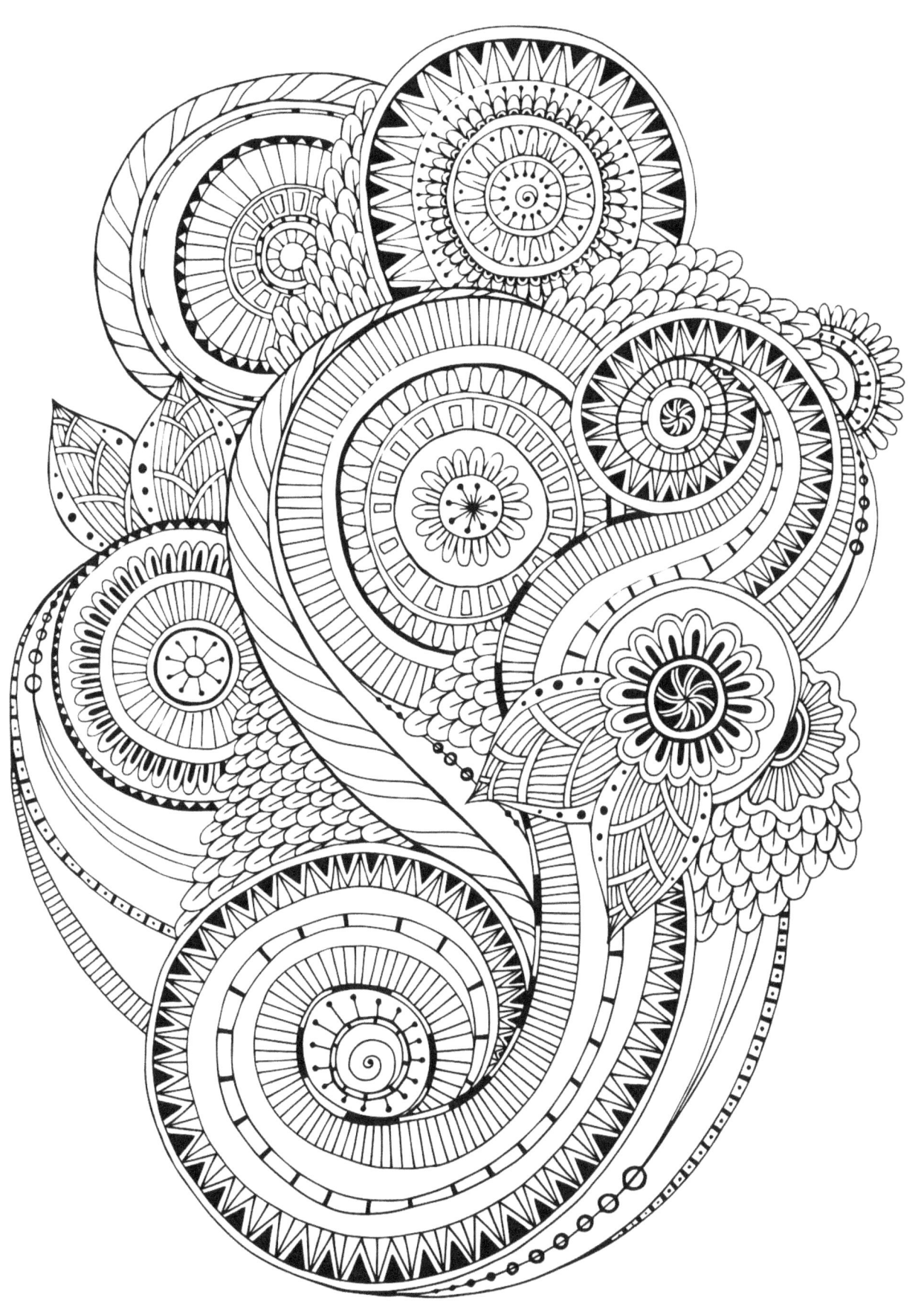

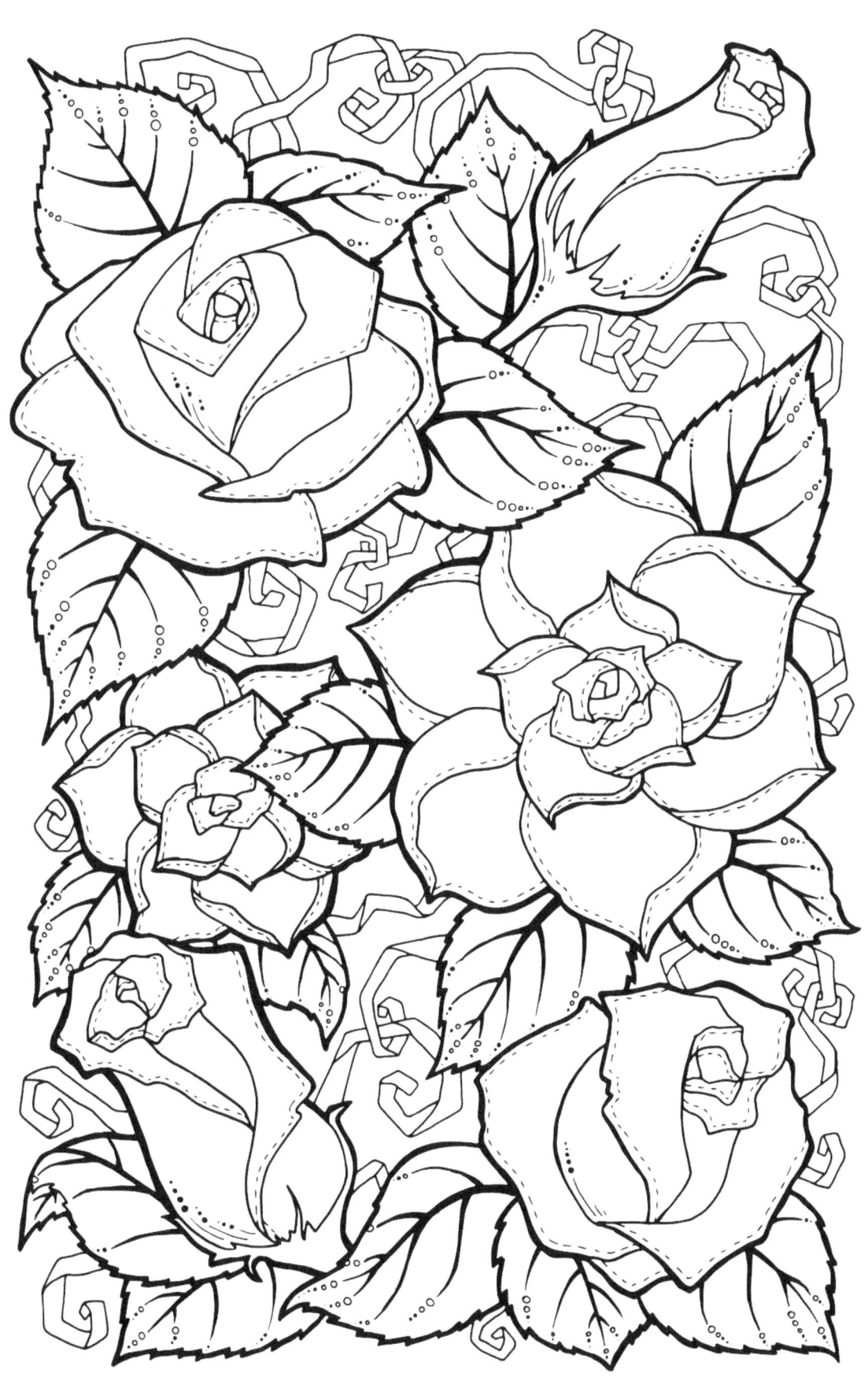

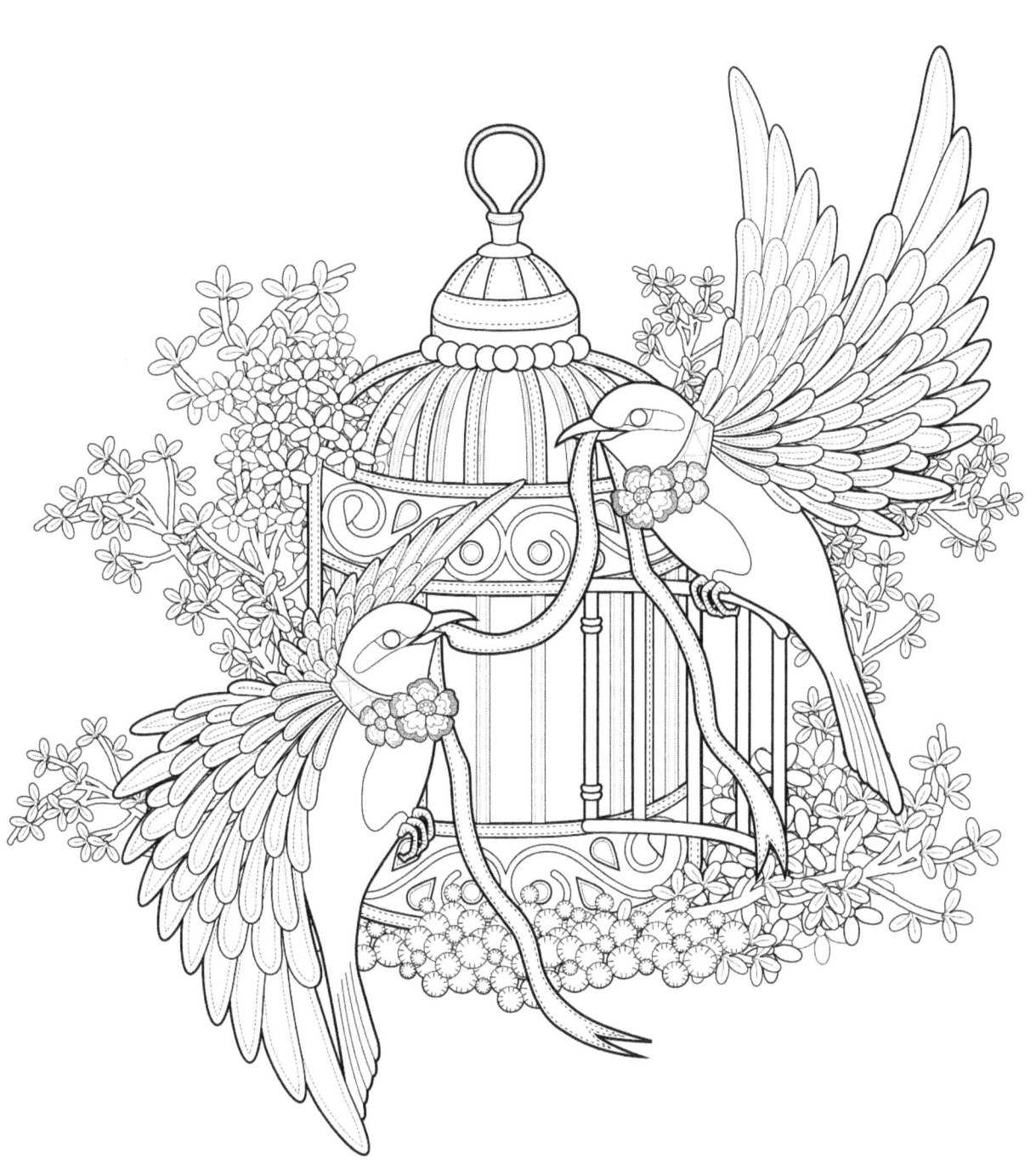

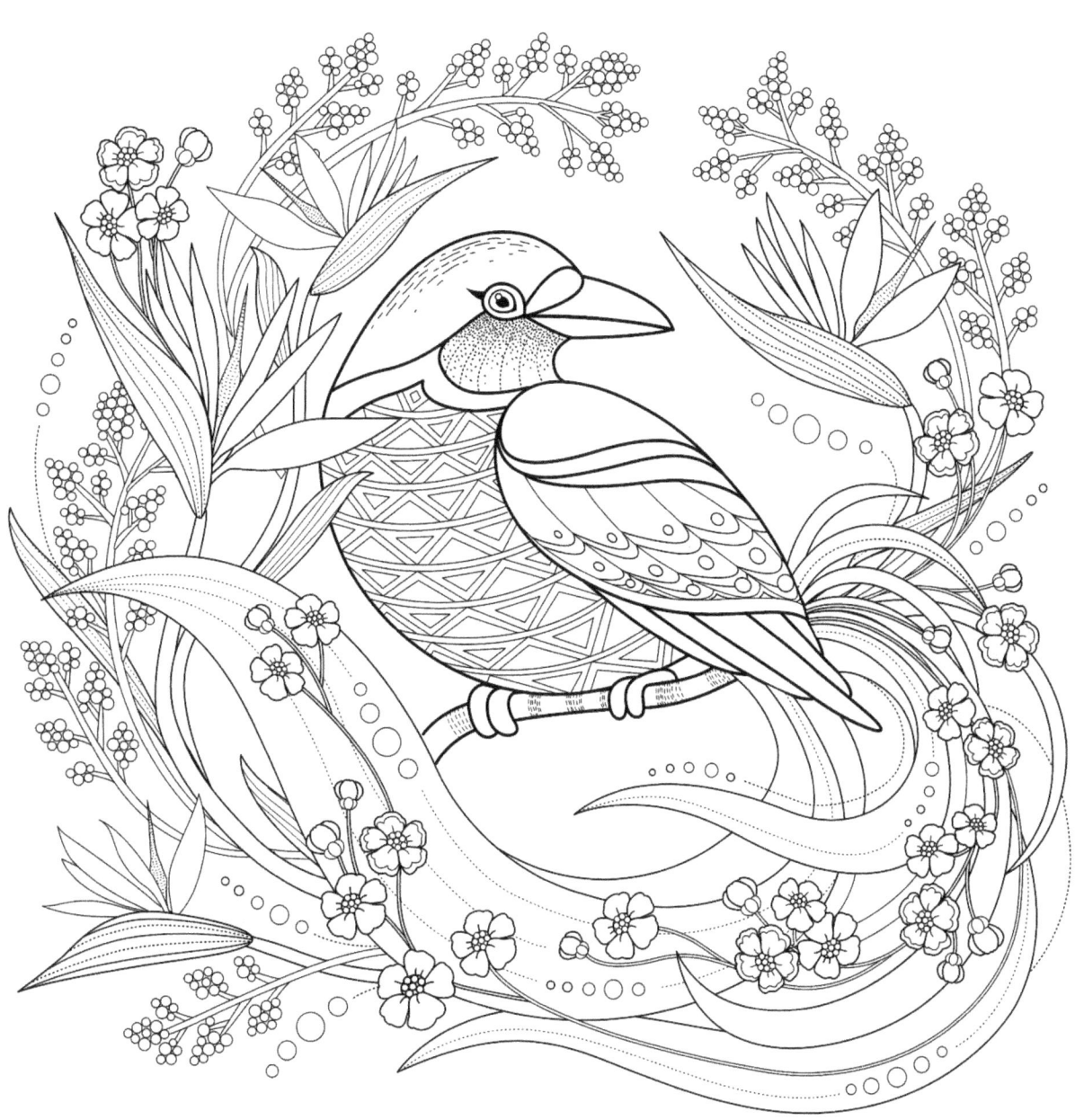

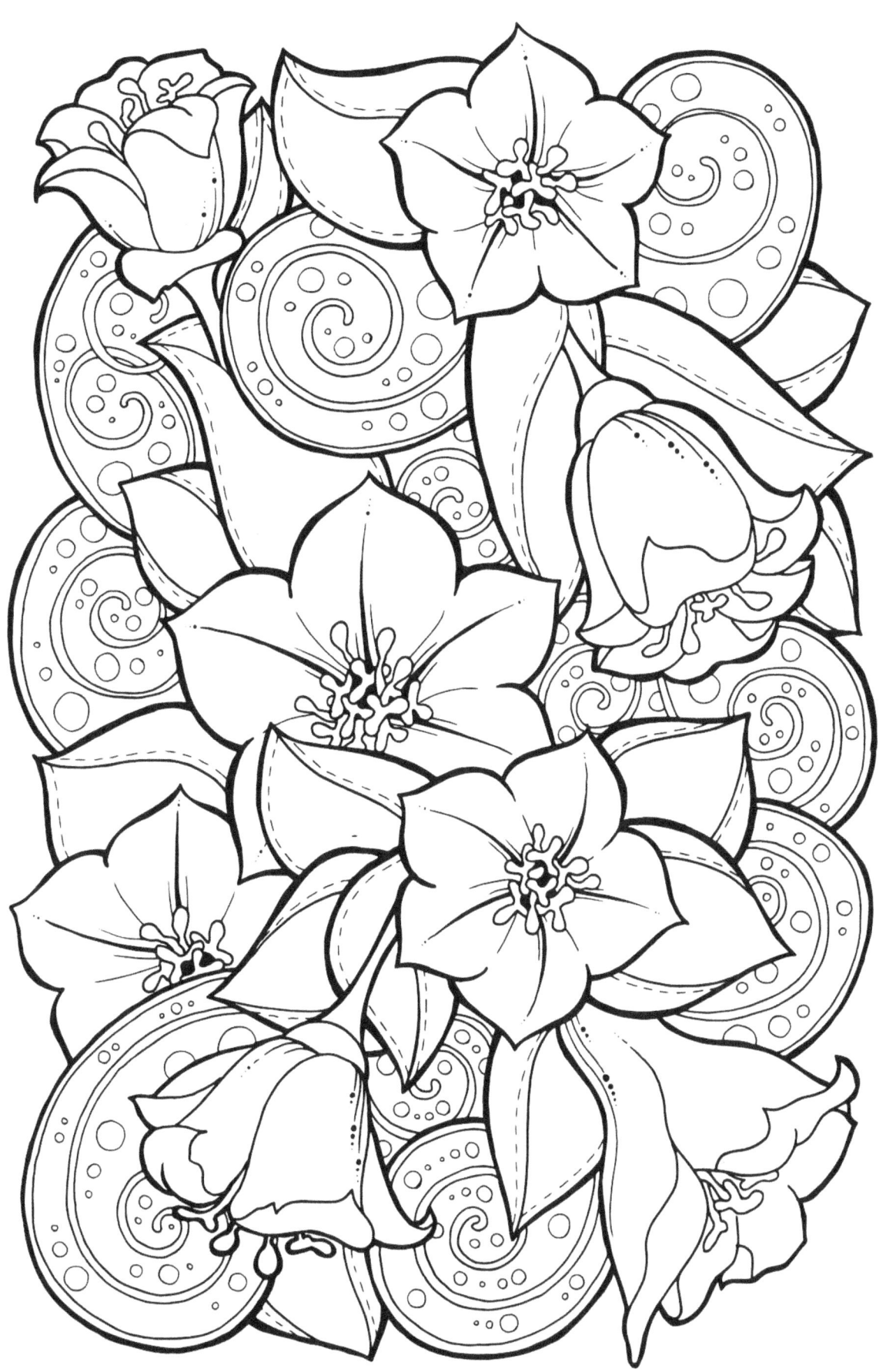

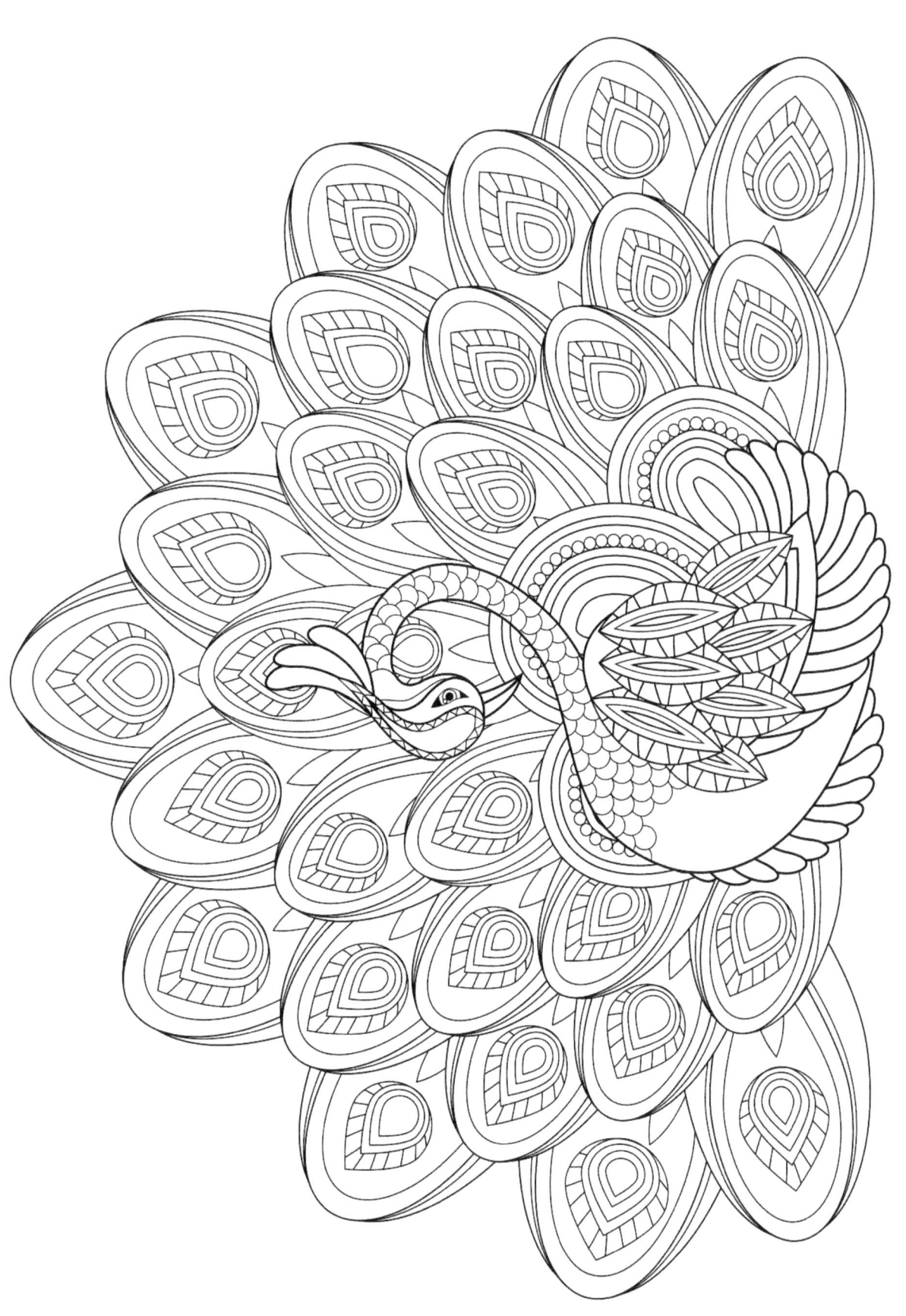

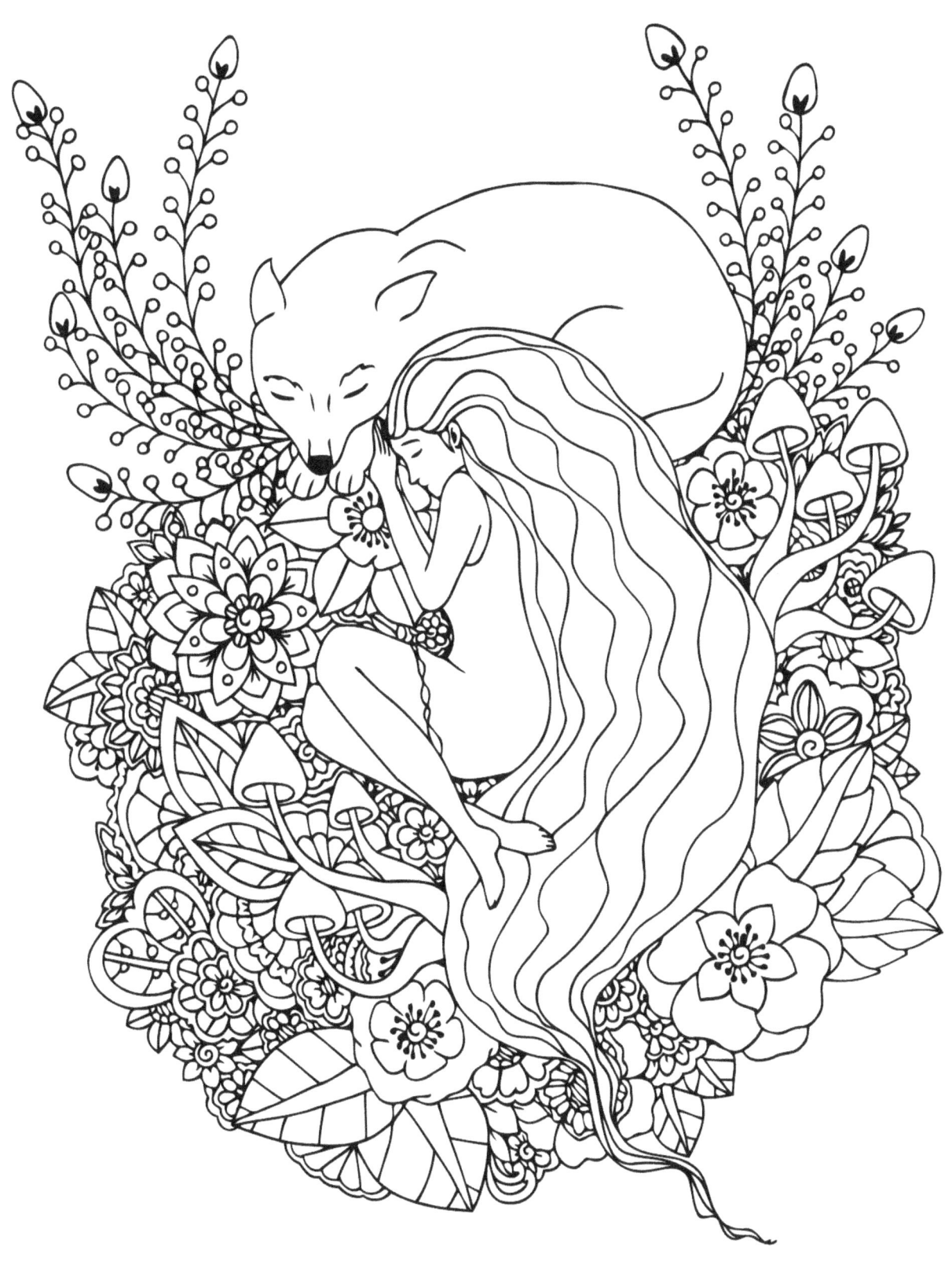

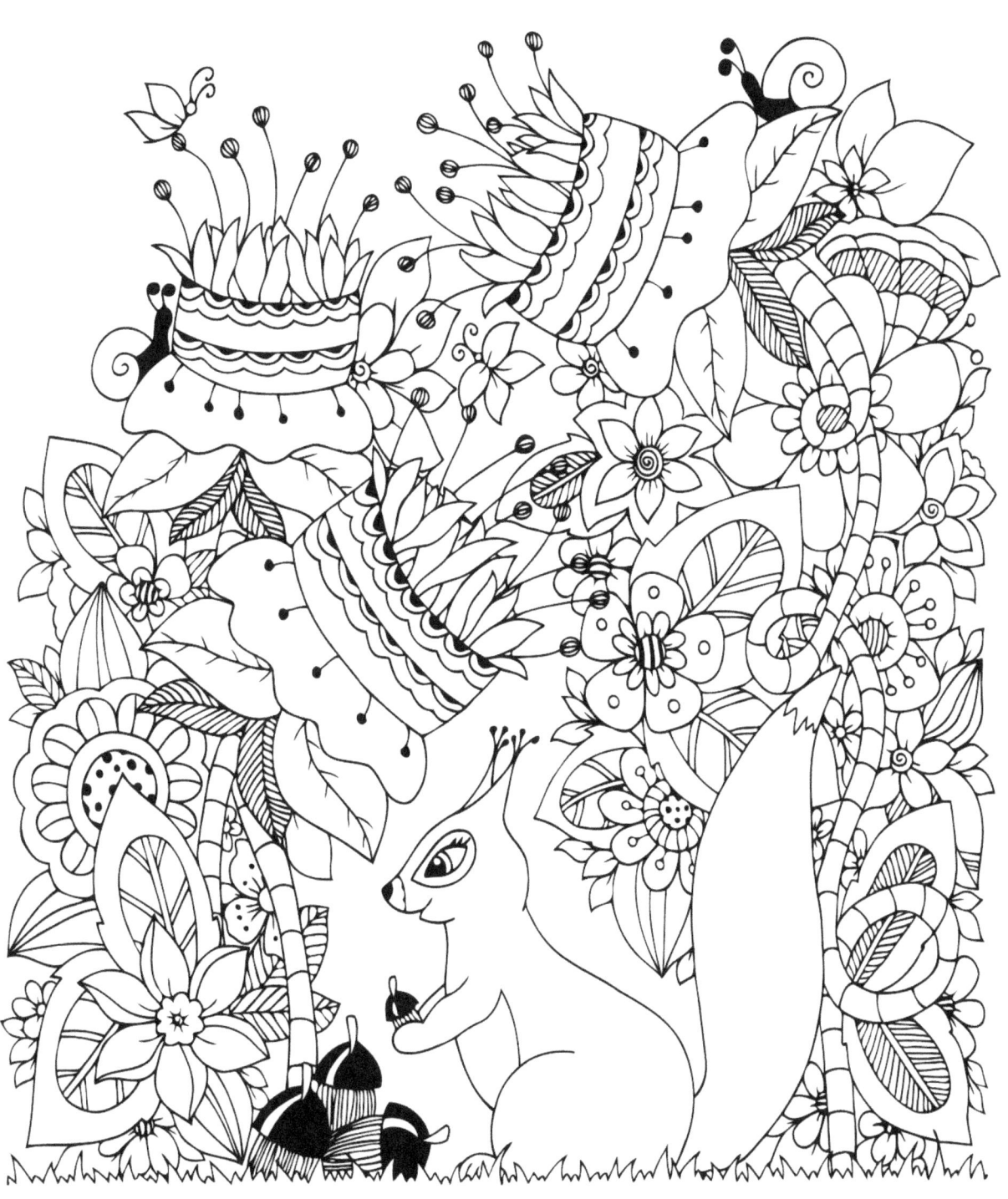

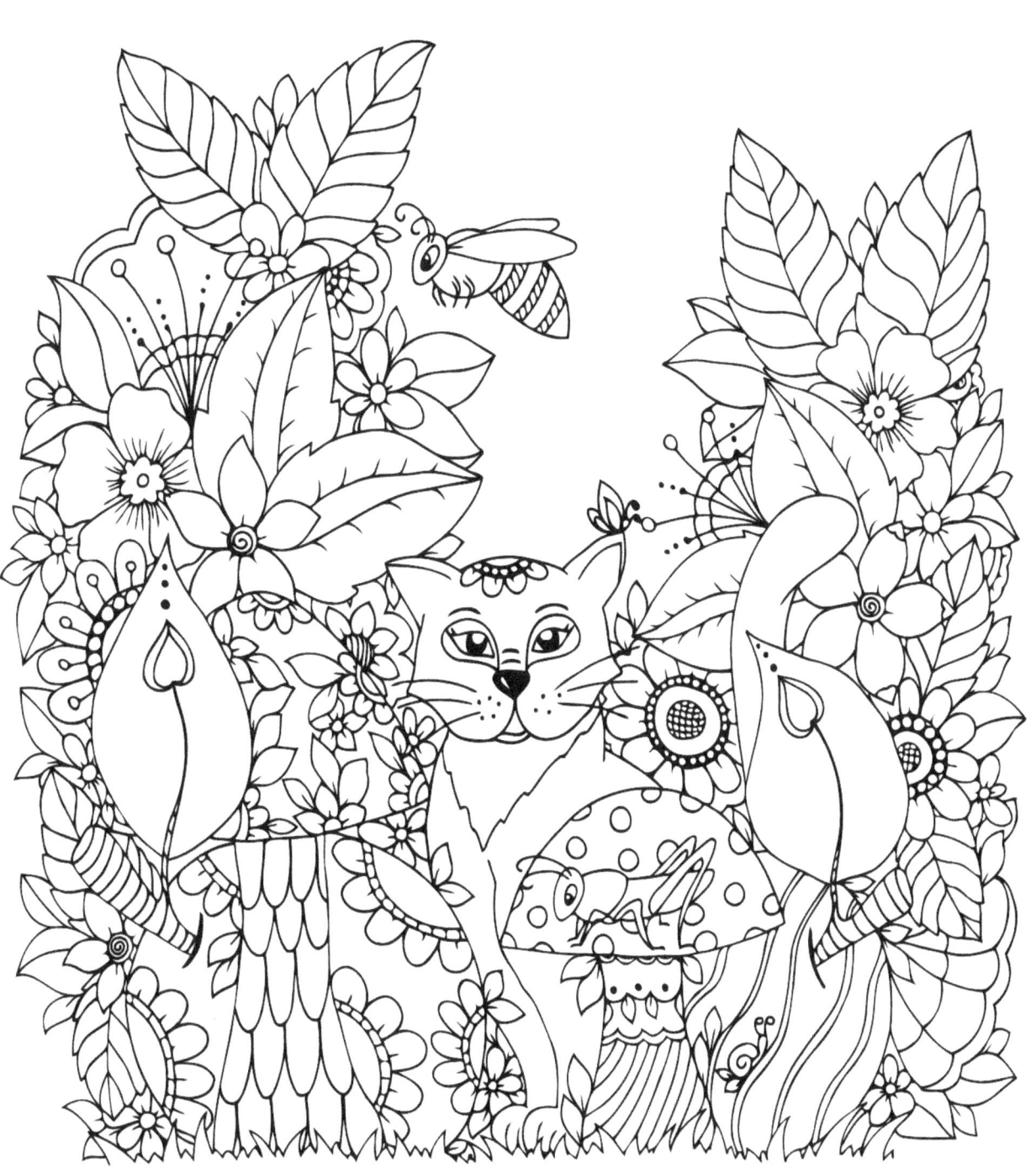

www.ingramcontent.com/pod-product-compliance
Lightning Source LLC
Chambersburg PA
CBHW081256180526
45170CB00007B/2446